Climate Change Debate

Series Editor: Cara Acred

Volume 286

Independence Educational Publishers

First published by Independence Educational Publishers
The Studio, High Green
Great Shelford
Cambridge CB22 5EG
England

British Library Cataloguing in Publication Data

Climate change debate. -- (Issues ; 286)
1. Global warming. 2. Climatic changes.
I. Series II. Acred, Cara editor.
363.7'3874-dc23

ISBN-13: 9781861687180

Printed in Great Britain

Zenith Print Group

Contents

Chapter 1: The climate crisis

What is climate change?	1
Climate change explained	2
Overview of greenhouse gases	6
How do scientists know that recent climate change is largely caused by human activities?	7
Climate change and the UK: risks and opportunities	8
Weird weather?	11
Climate change and nature	14
Warming oceans speeding up climate change cycle	15
Moving stories	16

Chapter 2: The climate debate

How does the IPCC know climate change is happening?	19
Who are the climate change deniers?	21
The real 'deniers' in the climate change debate are the warmists	22
Climate change: the science	23
Climate change scientists urged to be more open to the public about uncertainties	24
Climate is changing – but some believe the threat has been exaggerated	25
40% of adults worldwide have never heard of climate change	25
Britons believe in climate change... but do they care?	26

Chapter 3: Policies and solutions

Ten action areas for growth	27
UN and climate change: towards a climate agreement	30
Key points and questions: IPCC working group 3 report on mitigating climate change	32
UK pledges to help hardest hit by climate change	34
UN climate talks increasingly favour people alive today over future generations	35
Geoengineering: the ethical problems with cleaning the air	36
Fighting desertification will reduce the costs of climate change	37
How forestry helps address climate change	38
Why aren't more climate activists vegan?	38

Key facts	40
Glossary	41
Assignments	42
Index	43
Acknowledgements	44

Introduction

Climate Change Debate is Volume 286 in the ***ISSUES*** series. The aim of the series is to offer current, diverse information about important issues in our world, from a UK perspective.

ABOUT CLIMATE CHANGE DEBATE

39% of people believe that concerns about the world's climate changing have been exaggerated by scientists – should we be taking this climate crisis more seriously or is it being blown out of proportion? This book looks at the climate debate and considers the evidence for climate change and the risks involved, such as extreme weather events and migration issues. Topics also covered include how to mitigate climate change through legislation, geoengineering and even veganism.

OUR SOURCES

Titles in the ***ISSUES*** series are designed to function as educational resource books, providing a balanced overview of a specific subject.

The information in our books is comprised of facts, articles and opinions from many different sources, including:

- Newspaper reports and opinion pieces
- Website factsheets
- Magazine and journal articles
- Statistics and surveys
- Government reports
- Literature from special interest groups

A NOTE ON CRITICAL EVALUATION

Because the information reprinted here is from a number of different sources, readers should bear in mind the origin of the text and whether the source is likely to have a particular bias when presenting information (or when conducting their research). It is hoped that, as you read about the many aspects of the issues explored in this book, you will critically evaluate the information presented.

It is important that you decide whether you are being presented with facts or opinions. Does the writer give a biased or unbiased report? If an opinion is being expressed, do you agree with the writer? Is there potential bias to the 'facts' or statistics behind an article?

ASSIGNMENTS

In the back of this book, you will find a selection of assignments designed to help you engage with the articles you have been reading and to explore your own opinions. Some tasks will take longer than others and there is a mixture of design, writing and research-based activities that you can complete alone or in a group.

FURTHER RESEARCH

At the end of each article we have listed its source and a website that you can visit if you would like to conduct your own research. Please remember to critically evaluate any sources that you consult and consider whether the information you are viewing is accurate and unbiased.

Useful weblinks

www.2degreesnetwork.com

www.theccc.org.uk

www.climatemigration.org.uk

www.theconversation.com

www.eciu.net

http://energydesk.greenpeace.org

www.forestry.gov.uk

www.independent.co.uk

www.metoffice.gov.uk

naei.defra.gov.uk

www.newclimateeconomy.net

www.redd-monitor.org

www.royalsociety.org

www.rtcc.org

www.telegraph.co.uk

www.yougov.co.uk

The climate crisis

What is climate change?

Climate change is a large-scale, long-term shift in the planet's weather patterns or average temperatures. Earth has had tropical climates and ice ages many times in its 4.5 billion years. So what's happening now?

Since the last ice age, which ended about 11,000 years ago, Earth's climate has been relatively stable at about 14°C. However, in recent years, the average temperature has been increasing.

The information below details the seven main sources of evidence for climate change.

Higher temperatures

Scientific research shows that the climate – that is, the average temperature of the planet's surface – has risen by 0.89°C from 1901 to 2012. Compared with climate change patterns throughout Earth's history, the rate of temperature rise since the Industrial Revolution is extremely high.

Changes in nature

Changes in the seasons (such as the UK spring starting earlier, autumn starting later) are bringing changes in the behaviour of species; for example, butterflies appearing earlier in the year and birds shifting their migration patterns.

Sea ice

Arctic sea-ice has been declining since the late 1970s, reducing by about 4%, or 0.6 million square kilometres (an area about the size of Madagascar) per decade. At the same time Antarctic sea-ice has increased, but at a slower rate of about 1.5% per decade.

Changing rainfall

There have been observed changes in precipitation, but not all areas have data over long periods. Rainfall has increased in the mid-latitudes of the northern hemisphere since the beginning of the 20th century. There are also changes between seasons in different regions. For example, the UK's summer rainfall is decreasing on average, while winter rainfall is increasing. There is also evidence that heavy rainfall events have become more intensive, especially over North America.

Sea level rises

Since 1900, sea levels have risen by about 10cm around the UK and about 19cm globally, on average. The rate of sea-level rise has increased in recent decades.

Retreating glaciers

Glaciers all over the world – in the Alps, Rockies, Andes, Himalayas, Africa and Alaska – are melting and the rate of shrinkage has increased in recent decades.

Ice sheets

The Greenland and Antarctic ice sheets, which between them store the majority of the world's fresh water, are both shrinking at an accelerating rate.

⇨ The above information is reprinted with kind permission from the Met Office. Please visit www.metoffice.gov.uk for further information.

Climate change explained

Climate change is happening and is due to human activity, this includes global warming and greater risk of flooding, droughts and heat waves.

Causes of climate change

Rising levels of carbon dioxide and other gases, such as methane, in the atmosphere create a 'greenhouse effect', trapping the Sun's energy and causing the Earth, and in particular the oceans, to warm. Heating of the oceans accounts for over nine tenths of the trapped energy. Scientists have known about this greenhouse effect since the 19th century.

The higher the amounts of greenhouse gases in the atmosphere, the warmer the Earth becomes. Recent climate change is happening largely as a result of this warming, with smaller contributions from natural influences like variations in the Sun's output.

Carbon dioxide levels have increased by more than 40% since before the Industrial Revolution. Other greenhouse gases have increased by similarly large amounts. All the evidence shows that this increase in greenhouse gases is almost entirely due to human activity. The increase is mainly caused by:

- burning of fossil fuels for energy
- agriculture and deforestation
- the manufacture of cement, chemicals and metals

About 43% of the carbon dioxide produced goes into the atmosphere, and the rest is absorbed by plants and the oceans. Deforestation reduces the number of trees absorbing carbon dioxide and releases the carbon contained in those trees.

Evidence and analysis

Evidence from past climate change

Ancient ice from the polar ice sheets reveal natural temperature changes over tens to hundreds of thousands of years. They show that levels of greenhouse gases in the atmosphere are closely linked to global temperatures. Rises in temperature are accompanied by an increase in the amount of greenhouse gases.

These ice cores also show that, over the last 350 years, greenhouse gases have now rapidly increased to levels not seen for at least 800,000 years and very probably longer. Modern humans, who evolved about 200,000 years ago, have never previously experienced such high levels of greenhouse gases.

Natural fluctuations in climate

Over the last million years or so the Earth's climate has had a natural cycle of cold glacial and warm interglacial periods. This cycle is mainly driven by gradual changes in the Earth's orbit over many thousands of years, but is amplified by changes in greenhouse gases and other influences. Climate change is always happening naturally, but greenhouse gases produced by human activity are altering this cycle.

Volcanic eruptions and changes in solar activity also affect our climate, but they alone can't explain the changes in temperature seen over the last century.

Scientists have used sophisticated computer models to calculate how much human activity – as opposed to natural factors – is responsible for climate change. These models show a clear human 'fingerprint' on recent global warming.

Climate models and future global warming

We can understand a lot about the possible future effects of a warming climate by looking at changes that have already happened. But we can get much more insight by using mathematical models of the climate.

Climate models can range from a very simple set of mathematical equations (which could be solved using pen and paper) to the very complex, sophisticated models run on supercomputers (such as those at the Met Office).

While these models cannot provide very specific forecasts of what the weather will be like on a Tuesday in 100 years' time, they can forecast the big changes in global climate which we could see.

All these climate models tell us that by the end of this century, without an extremely significant reduction in the amount of greenhouse gases we produce, the world is likely to become more than 3°C warmer than in the 19th century. Note that this is a global average and that regional changes in some places will be even higher than this. There could even be global average rises of up to 6°C which would have catastrophic impacts.

This means that our action – or inaction – on greenhouse gas emissions today will have a substantial effect on climate change in the future.

The effects of climate change

We can already see the impacts of climate change and these will become more severe as global temperatures rise. How great the impacts will become depends upon our success in reducing greenhouse gas emissions.

The effects of rising temperatures on the UK

If global emissions are not reduced, average summer temperatures in the south east of England are projected to rise by:

- over 2°C by the 2040s (hotter than the 2003 heatwave which was connected to 2,000 extra deaths in the UK)
- up to nearly 4°C by the 2080s.

Rises in global temperature will have both direct and indirect effects on the UK. The UK's food supplies could be affected as crops in the UK and overseas could fail or be damaged by changes in temperature, rainfall and extreme weather events.

These extreme weather events in the UK are likely to increase with rising temperatures, causing:

- heavier rainfall events – with increased risk of flooding
- higher sea levels – with larger storm waves putting a strain on the UK's coastal defences
- more and longer-lasting heatwaves.

The effect of warming on rainfall patterns and water supplies

Changing rainfall patterns will affect water supplies. Too much rainfall in some areas and not enough in others will contribute to both flood and drought conditions. We are already seeing increasing numbers of heavy rainfall events, and expect this increase to continue, with greater risk of river and flash flooding.

Mountain glaciers are expected to continue melting which, along with reduced snow cover, will put stress on communities that rely on these as sources of water.

Changes in the oceans

Increasing temperatures and acidification of the oceans are threatening sea life around the world. Coral reefs in particular will be at major risk if ocean temperatures keep increasing.

Sea levels will keep rising as the polar ice sheets and glaciers melt and the warming oceans expand. Even small increases of tens of centimetres could put thousands of lives and properties at risk from coastal flooding during stormy weather.

Coastal cities with dense populations are particularly vulnerable, especially those that can't afford flood protection.

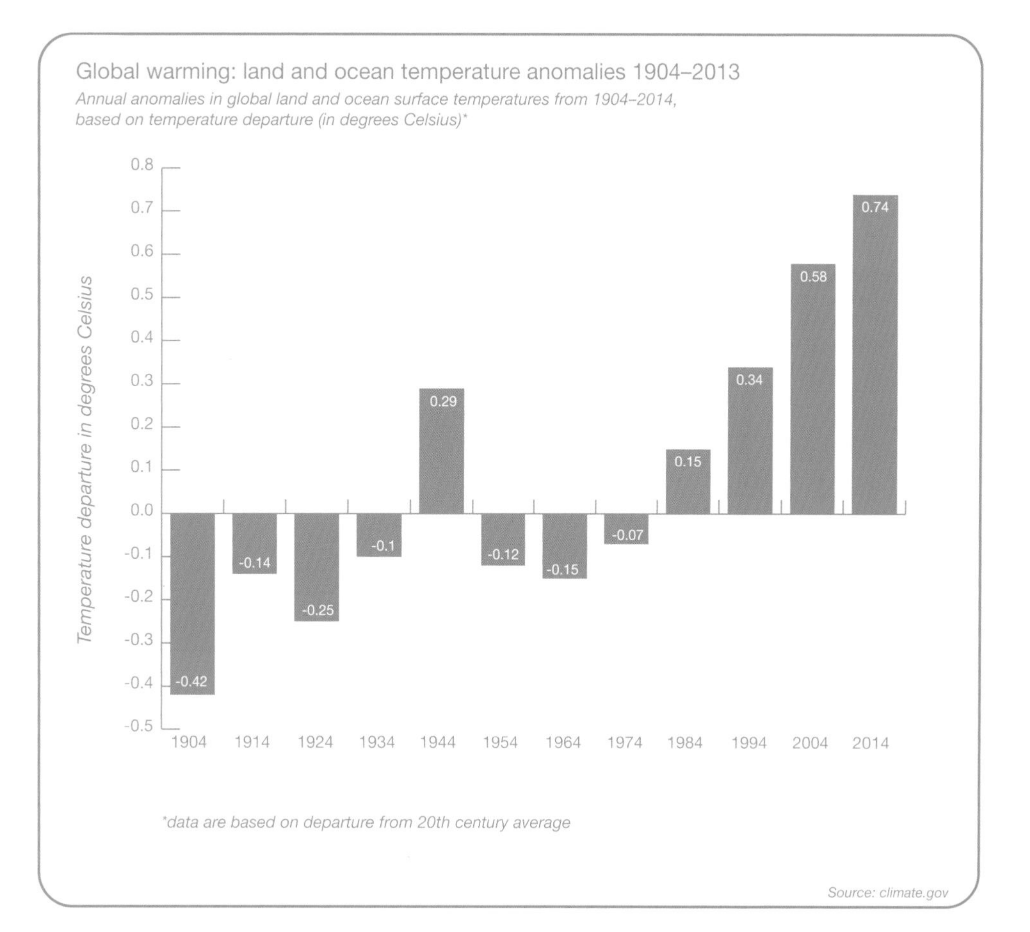

The impact of warming on food production

Even with low levels of warming (less than 2°C above the temperature in 1800), global production of major crops such as wheat, rice and maize may be harmed. Though warming may help some crops to grow better at high latitudes, food production in low latitudes will very likely suffer. This will cause a growing gap between food demand and supply.

Because trade networks are increasingly global, the effects of extreme weather events in one part of the world will affect food supply in another. For example, floods or droughts that damage crops in Eastern Europe or the US can directly affect the cost and availability of food in the UK.

The impact on ecosystems

Rapid, large changes in global temperatures (4°C or more above the temperature in 1800 by the end of this century) could cause the extinction of entire species. Even with smaller amounts of warming, species will be placed more at risk. The animals and plants most at risk will be those that:

- have no new habitats to move to
- can't move quickly to new habitats
- are already under threat from other factors.

Extinctions could have an enormous impact on the food chain. Most ecosystems would struggle to live with large changes in climate which happen rapidly within a century or so.

The impact on human health

Climate change is expected to make some existing health problems worse as temperatures increase. Malnutrition could become more widespread as crops are affected and warmer temperatures could increase the range of disease-carrying insects. Vulnerable people will be at risk of increased heat exposure, although there will likely be fewer health problems related to cold temperatures.

Poverty

Populations with low income in both developed and developing countries will be most vulnerable to the impacts of climate change. Decreasing food production, an increase in health issues associated with climate change, and more extreme weather will slow economic growth, making it increasingly difficult to reduce poverty.

The impact of extreme weather events globally

Growing populations and increasingly expensive infrastructure are making our societies more vulnerable to extreme weather events. Heatwaves and droughts are expected to become more common and more intense over the coming century, and more frequent heavy rainfall events and rising sea levels will increase the risk of floods.

While not all extreme weather events can be directly linked to human influences, we are already seeing the huge impacts on society that extreme weather events can have. The World Meteorological Organization (WMO) reported that between 2001 and 2010 extreme weather events caused:

- more than 370,000 deaths worldwide (including a large increase in heatwave deaths from 6,000 to 136,000) – 20% higher than the previous decade
- an estimated US$660 billion of economic damage – 54% higher than in the previous decade.

Possible abrupt changes in our climate

Most discussions of climate change look at what is most likely to happen, such as the likely temperature changes if we do, or don't, take action to reduce greenhouse gas emissions.

But scientists have identified the possibility that with sustained high temperatures major elements of the Earth's climate could be drastically altered. These 'tipping points' in our climate are less likely, but potentially much more dangerous.

While known impacts from small temperature rises could be managed (although this will become increasingly expensive as temperatures increase), passing a tipping point could cause large or abrupt changes, some of which may be effectively irreversible.

For example:

- Arctic permafrost could thaw rapidly, releasing greenhouse gases that are currently 'locked away' and causing further rapid warming
- the great sheet of ice covering Greenland, which contains enough ice to cause up to seven metres of sea level rise, could almost entirely melt. While this would take a long time to happen, it is possible that the ice sheet would not be able to regrow after a certain amount of melting occurs.

While such events are considered unlikely, they can't be ruled out, even under relatively low temperature rises of less than 2°C above the temperature in 1800. All indications are that, should we pass one of these tipping points, there would be a range of extremely severe impacts.

Agreement among experts

Overwhelming amounts of scientific evidence show that the planet is warming and that human activity is the main contributor to this warming.

Many leading national scientific organisations have published statements confirming the need to take action to prevent potentially dangerous climate change. These include:

- The G8+5 National Science Academies' Joint National Statement which represents the UK, along with Brazil, Canada, China, France, Germany, India, Italy, Japan, Russia and the United States
- The American Association for the Advancement of Science (AAAS) statement
- The Royal Society and US National Academy of Sciences have produced an authoritative and accessible report on *Climate Change Evidence and Causes* which provides answers to many common questions.

You can find out more about the scientific evidence on climate change from:

- The Met Office Hadley Centre
- Frequently Asked Questions from the Intergovernmental Panel on Climate Change
- The UK Geological Society.

The role of the IPCC

The Intergovernmental Panel on Climate Change (IPCC) is an independent body composed of scientists from around the world. It that has been tasked by the United Nations to assess and review the most recent scientific, technical and socio-economic evidence related to climate change.

The IPCC's fifth assessment science report concluded that the scientific evidence for a warming climate is undeniable and that 'human influence on the climate system is clear'.

The UK Government has always fully supported the work of the IPCC and regards its assessments as the most authoritative view on the science of climate change available.

DECC's summaries of the IPCC fifth Assessment reports 2013/14

- *The Physical Science Basis of Climate Change report*, 27 September 2013
- *Impacts, Adaptation and Vulnerability* report, 31 March 2014
- *Mitigation of Climate Change* report, 12 April 2014.

Tackling climate change

If we take action to radically reduce greenhouse gas emissions now, there's a good chance that we can limit average global temperature rises to 2°C. This doesn't mean that there will be no more changes in the climate – warming is already happening – but we could limit, adapt to and manage these changes.

If we take action now:

- we will avoid burdening future generations with greater impacts and costs of climate change
- economies will be able to cope better by mitigating environmental risks and improving energy efficiency
- there will be wider benefits to health, energy security and biodiversity.

The economic benefit of taking action now

It makes good economic sense to take action now to drastically cut greenhouse gas emissions. If we delay acting on emissions, it will only mean more radical intervention in the future at greater cost.

Taking action now can also help to achieve long-term, sustainable economic growth from a low-carbon economy.

UK Government action

The UK Government is:

- working to secure global emissions reductions
- reducing UK emissions
- adapting to climate change in the UK.

23 October 2014

- The above information is reprinted with kind permission from the Department of Energy & Climate Change. Please visit www.gov.uk for further information.

Overview of greenhouse gases

Introduction

The GHG inventory covers the seven direct greenhouse gases under the Kyoto Protocol:

- Carbon dioxide (CO_2)
- Methane (CH_4)
- Nitrous oxide (N_2O)
- Hydrofluorocarbons (HFCs)
- Perfluorocarbons (PFCs)
- Sulphur hexafluoride (SF_6)
- Nitrogen trifluoride (NF_3)

These gases contribute directly to climate change owing to their positive radiative forcing effect. HFCs, PFCs, SF_6 and NF_3 are collectively known as the 'F-gases'.

In general terms, the largest contributor to global warming is carbon dioxide which makes it the focus of many climate change initiatives. Methane and nitrous oxide contribute to a smaller proportion, typically <10%, and the contribution of F-gases is even smaller (in spite of their high Global Warming Potentials) at <5% of the total.

Also reported are four indirect greenhouse gases:

- Nitrogen oxides
- Carbon monoxide
- Non-methane volatile organic compounds (NMVOC)
- Sulphur dioxide.

Nitrogen oxides, carbon monoxide and NMVOCs are included in the inventory because they can produce increases in tropospheric ozone concentrations and this increases radiative forcing (warming of the atmosphere). Sulphur dioxide is included because it contributes to aerosol formation which can either warm (through absorption of solar radiation on dark particles) or cool (from forming cloud droplets and reflecting radiation) the atmosphere.

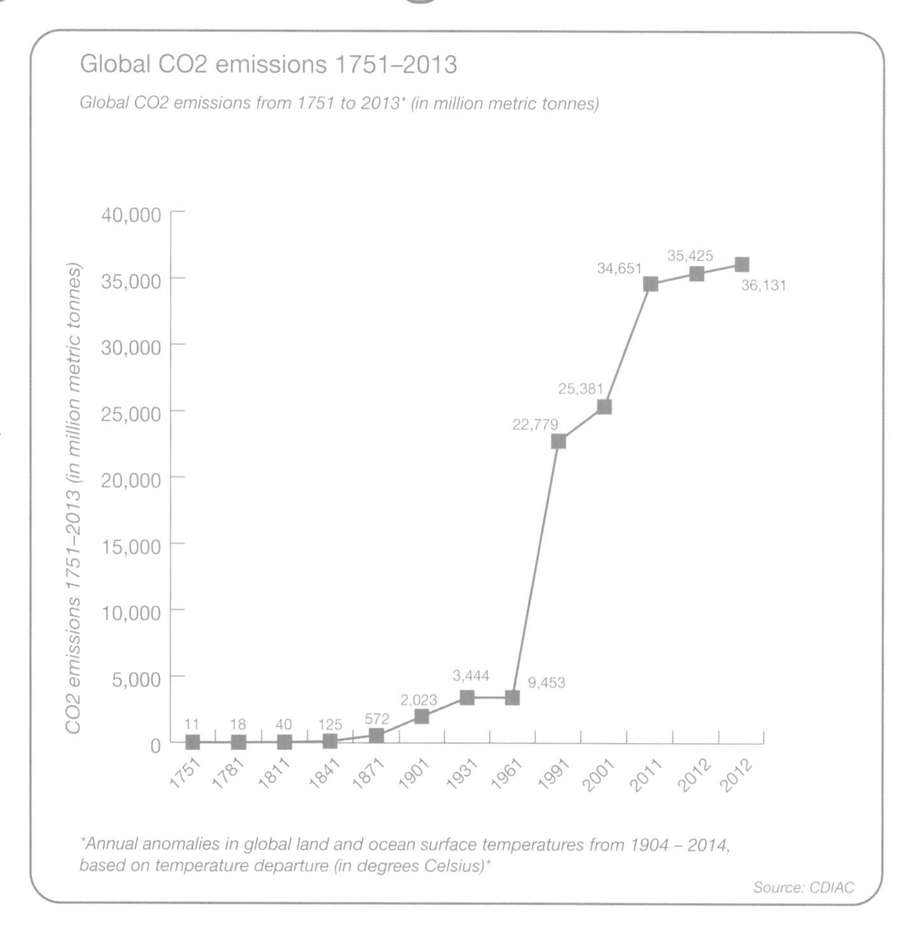

Important sources and trends

The largest source is the combustion of fossil fuels in the energy sector, typically approximately 85% of total emissions. Emissions of CO_2, CH_4 and N_2O arise from this sector. Energy sector emissions have declined since the base year of 1990, primarily due to fuel switching to less carbon-intensive energy sources (e.g. coal to gas in the power sector) and reduced energy intensity of the economy (e.g. declining iron and steel and metal production industries).

The second largest source of greenhouse gases in the UK is the agricultural sector at up to 10% of total emissions. Emissions from this sector arise for both CH_4 and N_2O. Since 1990, emissions from this sector have declined due to a reduction in livestock numbers, changes in the management of manure and restrictions in the use of synthetic fertiliser.

Industrial processes sector makes up the third largest source of greenhouse gases in the UK, contributing to up to 5% of the national total. Emissions of all six direct greenhouse gases occur from this sector. Emissions from this sector include non–energy related emissions from the production and use of cement and lime, chemical industry and metal production as well as F-gases from refrigeration, air conditioning and other industrial and product use. Since 1990, this category has seen a large decline in emissions, mostly due to a reduction in bulk chemical production and metal processing industries and due to changes in process.

The Land Use, Land Use Change and Forestry (LULUCF) sector contains absorbers (sinks) as well as sources of CO_2 emissions. LULUCF has been a net sink since 2001. Emissions from this source occur for CO_2, N_2O and CH_4 from clearing of forests and vegetation, flooding of land and from application of fertilisers and lime.

Emissions from the waste sector contributed <3% to greenhouse gas emission. The majority from CH4 from solid waste disposal on land. Overall emissions from the waste sector have decreased since 1990, mostly due to the implementation of methane recovery systems at UK landfill sites, and reductions in the amount of waste disposed of at landfill sites.

Targets

Countries that have signed and ratified the Kyoto Protocol are legally bound to reduce their greenhouse gas emissions by an agreed amount. A single European Union Kyoto Protocol reduction target for greenhouse gas emissions of -8% compared to base-year levels was negotiated for the first commitment period, and a Burden Sharing Agreement allocated the target between Member States of the European Union. Under this agreement, the UK reduction target was -12.5% on base-year levels. The first commitment period of the Kyoto Protocol was from 2008 to 2012.

The second commitment period of the Kyoto Protocol will run for eight years, from 2013 to 2020 inclusive. For this second commitment period, alongside the EU and its member States, the UK (including Gibraltar) communicated an independent quantified economy-wide emission reduction target of a 20% emission reduction by 2020 compared with 1990 levels (base year). The target for the European Union and its Member States is based on the understanding that it will be fulfilled jointly with the European Union and its Member States. The 20% emission reduction target by 2020 is unconditional and supported by legislation in place since 2009 (Climate and Energy Package). Once ratified this Kyoto target will cover the UK, and the relevant Crown Dependencies and Overseas Territories that wish to join the UK's ratification. As ratification is not yet complete, the exact details of the UK's target for the second commitment period are still being finalised.

The Climate Change Act became UK Law on 26 November 2008. This legislation introduced ambitious and legally binding national targets for the UK to reduce GHG emissions to 34% below base year by 2020 and to 80% below base year by 2050. These targets are underpinned with legally binding five-year GHG budgets.

Further information on the UK's action to tackle climate change can be found on the DECC website: https://www.gov.uk/government/organisations/department-of-energy-climate-change

13 August 2015

⇨ The above information is reprinted with kind permission from National Atmospheric Emission Inventory. Please visit naei.defra.gov.uk for further information.

How do scientists know that recent climate change is largely caused by human activities?

Scientists know that recent climate change is largely caused by human activities from an understanding of basic physics, comparing observations with models, and fingerprinting the detailed patterns of climate change caused by different human and natural influences.

Since the mid-1800s, scientists have known that CO_2 is one of the main greenhouse gases of importance to Earth's energy balance. Direct measurements of CO_2 in the atmosphere and in air trapped in ice show that atmospheric CO_2 increased by about 40% from 1800 to 2012. Measurements of different forms of carbon (isotopes) reveal that this increase is due to human activities. Other greenhouse gases (notably methane and nitrous oxide) are also increasing as a consequence of human activities. The observed global surface temperature rise since 1900 is consistent with detailed calculations of the impacts of the observed increase in atmospheric CO_2 (and other human-induced changes) on Earth's energy balance.

Different influences on climate have different signatures in climate records. These unique fingerprints are easier to see by probing beyond a single number (such as the average temperature of Earth's surface), and looking instead at the geographical and seasonal patterns of climate change. The observed patterns of surface warming, temperature changes through the atmosphere, increases in ocean heat content, increases in atmospheric moisture, sea level rise, and increased melting of land and sea ice also match the patterns scientists expect to see due to rising levels of CO_2 and other human-induced.

The expected changes in climate are based on our understanding of how greenhouse gases trap heat. Both this fundamental understanding of the physics of greenhouse gases and fingerprint studies show that natural causes alone are inadequate to explain the recent observed changes in climate. Natural causes include variations in the Sun's output and in Earth's orbit around the Sun, volcanic eruptions, and internal fluctuations in the climate system (such as El Niño and La Niña). Calculations using climate models have been used to simulate what would have happened to global temperatures if only natural factors were influencing the climate system. These simulations yield little warming, or even a slight cooling, over the 20th century. Only when models include human influences on the composition of the atmosphere are the resulting temperature changes consistent with observed changes.

7 February 2014

⇨ The above information is reprinted with kind permission The Royal Society. Please visit www.royalsociety.org for further information.

Climate change and the UK: risks and opportunities

What does the Intergovernmental Panel on Change Fifth Assessment Report mean for the UK?

Science

- Climate change is unequivocal, with evidence observed on every continent and in the ocean. It is at least 95% likely that human agency, particularly the emission of greenhouse gases, is the dominant cause.
- The average temperature at the Earth's surface has risen faster over the past century than at any time in the past two millennia, and probably much longer. The rate of increase has slowed over the last 16 years but other indicators of climate change such as ocean warming, sea level rise and melting glaciers have not slowed – nor has acidification of the ocean.
- The concentration of carbon dioxide in the atmosphere is higher than it has been for at least 800,000 years. If emissions continue rising at the current rate, consequences by the end of the century include a global average temperature 2.6–4.8°C higher than present and sea level 0.45–0.82 metres higher. This warming presents risks of triggering major and irreversible changes such as melting of the Greenland ice sheet.
- Projecting changes for the UK is not straightforward given our geographical position, with weather influenced by changeable factors including the jet stream and the Gulf Stream. However, an increase in extreme events such as heatwaves, increasing flood risk due to rising sea level and rain, and increased drought risk in the south appear likely.
- Projected global trends with impacts on Britons include a reduction in agricultural output, and damage to nature including increased risk of species extinctions, disruption of the marine food web and increased pressures for species migration

'The overall scientific picture of climate change is clearer than it has ever been, although our understanding will never be perfect. What we do know is that the climate system takes decades to respond to changes in carbon emissions, so decisions made now will affect future generations as much as the current one.' – *Professor Joanna Haigh Co-Director, Grantham Institute, Imperial College London*

Economics, adaptation and mitigation

- Investing in adaptation is prudent because greenhouse gases already in the system make further climate impacts inevitable. Many aspects of climate change, and ocean acidification, will continue for centuries even if emissions fall rapidly.
- Capacity to adapt is progressively eroded as climate impacts progress. A combination of adaptation measures and substantial cuts in greenhouse gas emissions can limit risks from climate change.
- Limiting global warming to 2° Celsius, the internationally agreed goal, is feasible if governments act quickly and in concert. Most feasible pathways have emissions peaking within about a decade and declining to zero in the second half of the century. Costs of the transition to 2°C would shave about 0.06% off annual economic growth if governments enacted co-ordinated measures such as a global carbon price.
- The sum of current pledges from governments is not enough to meet the 2°C target. On current trajectories, 3°C or 4°C is more likely. Few studies have mapped the economic consequences of such a change.

- Transition to the 2°C pathway implies rapid and profound changes in areas such as energy supply, buildings and transport. Early transition in the energy sector is economically and technically favoured: measures include substantial energy efficiency improvements, a three-to-four fold increase in low-carbon generation by 2050, and electrification of services such as heating and transport.

'The evidence clearly shows that a prudent climate change policy involves both adaptation and decarbonisation; neither on its own is enough. The evidence suggests that decarbonisation will be cheaper if it starts earlier, with reducing energy waste a key priority.' – *Professor Jim Skea Research Councils UK Energy Strategy Fellow and Professor of Sustainable Energy, Imperial College London: member of the Committee on Climate Change: Vice-Chair, Working Group 3, Intergovernmental Panel on Climate Change*

Health

- The risk of extreme heat events (heatwaves) is likely to have increased in large parts of Europe due to climate change. Further extremes are expected in future with the very young and the elderly most vulnerable to the effects of heat.
- There could be modest reductions in cold-related mortality and morbidity due to fewer cold extremes, although cold snaps will still occur.
- More flooding would increase the risk of deaths from drowning, the spread of infectious diseases and risks of mental health conditions. The prevalence of mental health problems (distress, anxiety, and depression) was two to five times higher among individuals who reported flooding in the home in 2007.
- New and resurgent vector-borne diseases, including Dengue, are moving north as temperatures rise, and have recently appeared in Europe.
- Overall, the impact to health from climate change in the UK is likely to be negative, but tackling climate change can have significant 'co-benefits' for health. These include reduced air pollution from emissions, and improving general fitness and cardiovascular health through shifting to 'active transport' such as cycling and walking.

'Climate change is a health issue, and we take climate change very seriously. The science is clear and the time to act is now. We anticipate significant benefits to health from a low-carbon lifestyle, and significant impacts on health if we don't.' – *Professor David Walker Deputy Chief Medical Officer for England*

Food and farming

- Extreme weather conditions are projected to increase and are likely to hit UK food production. (Extreme weather led to the UK needing to import wheat in 2012 and 2013.)
- Moderate warming may expand the growing season of some crops in the UK. But the overall impact on crop yields is unclear.
- Globally, climate change is likely to reduce crop yields over time. Given the needs of an expanding global population, the security of supply of our imported food and livestock feed is at risk, with more volatile prices likely.
- Agriculture needs land and water, but both are threatened by more droughts and floods. (Farmers will need access to enough water and, with most of the UK's best land below the 5m contour line and sea levels rising, more investment in flood management is needed.)

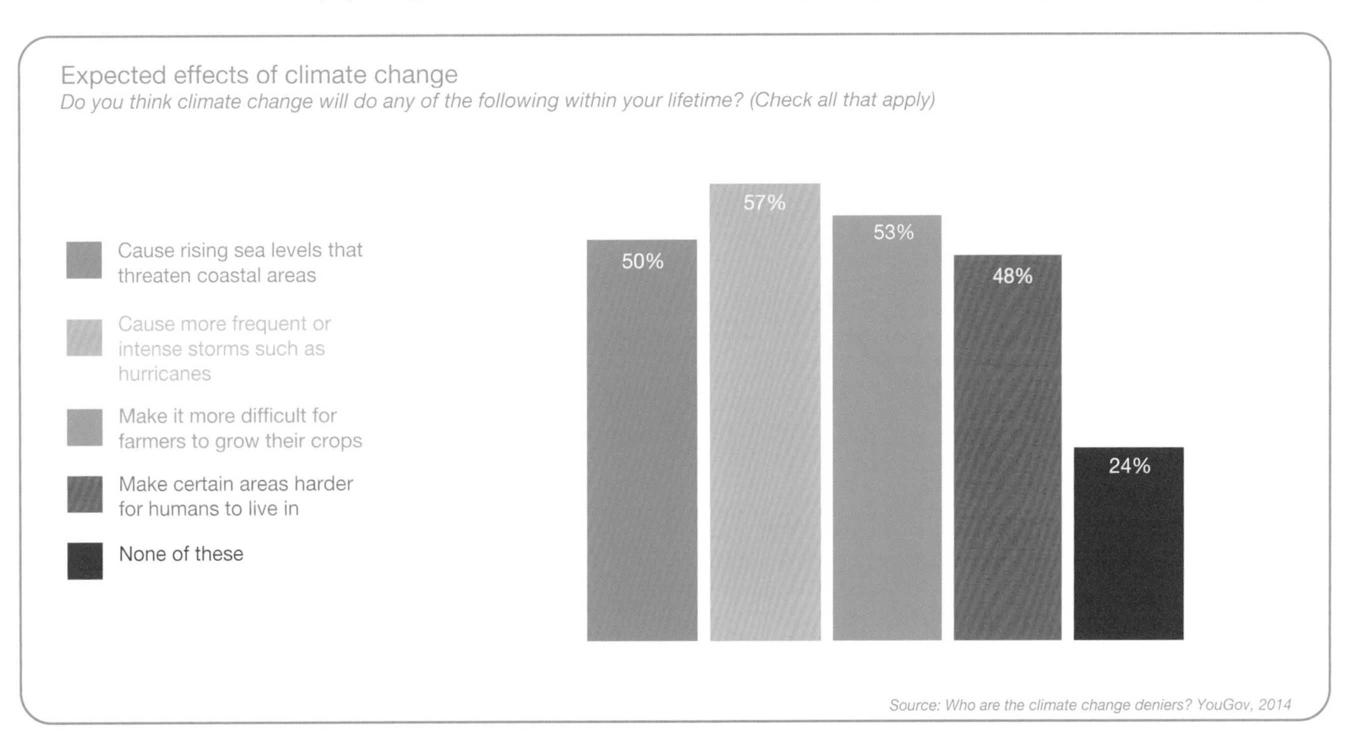

- Heat stress also reduces yields in livestock, whilst the gradual warming already seen has led to the arrival of new diseases such as Bluetongue.

'The IPCC's review of the latest evidence points to the range of possible challenges ahead. The impacts on agriculture of the extreme events this decade alone mean that we can't keep on gambling on our ability to produce food in the UK whatever the weather.' *– Guy Smith Vice-President, National Farmers' Union*

Security and the military

- Climate change poses an increasing risk to geopolitical stability and therefore the security and well-being of all.
- Climate change acts as a 'threat multiplier', exacerbating existing threats to peace and security. This is likely to result in an increased demand for UK military engagement, be it in the form of conflict prevention, conflict resolution or humanitarian assistance
- Many developing countries, including some in the Commonwealth, are already feeling impacts of a changing climate, in addition to stresses such as poverty and shortages of food and water. Some governments are already struggling to meet the needs of their populations through lack of capacity and resilience.
- Increased instability in countries affected by climate change poses a risk to supplies of essential raw materials and to the emerging markets that the UK requires for sustained economic growth.
- There is no security solution to climate change, just a greater risk of insecurity if we choose not to act.

'The UK military, along with colleagues around the world, recognises the threat that climate change poses to global security and our national interests. They will act to address the challenges this presents, but they cannot and should not be doing it on their own. As the IPCC concludes, limiting the risks of climate change requires a combination of adaptation measures and substantial cuts in greenhouse gas emissions.' – *Rear-Admiral Neil Morisetti Former Commander of UK Maritime Forces and UK Climate and Energy Security Envoy. Currently Director of Strategy, Department of Science, Technology, Engineering and Public Policy, University College London*

Business

- Climate change presents a variety of risks and opportunities to UK businesses; there is no 'one size fits all' implication. Risks and opportunities come from climate impacts and climate policies.

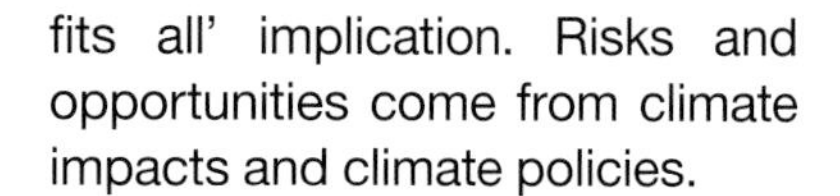

- Sectors likely to be adversely affected by direct climate impacts include food, tourism, fisheries and insurance. Sectors likely to be affected by climate policy include energy, construction, transport and manufacturing. All sectors are likely to experience increasing disruption to supply chains.
- In general, businesses that adopt proactive strategies are likely to thrive. Such strategies include supply chain resilience, anticipation of and adaptation to direct climate impacts, reducing energy consumption, innovating and investing in low-carbon technologies. Failing to prepare means preparing to fail.
- For investors, some traditional blue-chip stocks such as fossil fuel companies will become increasingly risky. Gas providers could benefit temporarily from a low-carbon transition, but coal companies will not.
- Businesses have an interest in encouraging governments to adapt and reduce emissions quickly, as modelling shows the economic costs of constraining climate change at any given level are higher if action is deferred.

'The opportunities over the next 15 years are tremendous – but so are the risks. Overall, the best strategies for businesses are rooted in understanding the risks that climate change brings, and the opportunities offered by the transition to a resilient, low-carbon business model.' – *Jeremy Oppenheim Director, McKinsey & Co; Programme Director, New Climate Economy*

2014

- The above information is reprinted with kind permission from the Energy and Climate Intelligence Unit. Please visit www.eciu.net for further information.

Weird weather?

Four weather events that could be linked to climate change.

By Christine Ottery, EU Energydesk editor

You've probably noticed there's been some weird stuff going on with the weather lately.

The sheer amount of different extreme weather events going on simultaneously around the world means this could be the winter when climate change becomes 'real' in our minds, after more than two decades of scientists telling us what its impacts would be.

The recent IPCC AR5 report concluded the climate is changing and there is a 95% certainty that it is caused by our actions – specifically the burning of fossil fuels, deforestation and land use change.

But the World Meteorological Organization (WMO) told Energydesk: 'No single weather episode can prove or disprove global climate change.' Right, and of course many complicated and interrelated variables are in play, of which climate change might only be one factor.

And yes, and there are problems with attributing specific weather to climate change until more sophisticated reverse modeling is possible – an issue examined in a 2013 modelling paper identifying the contribution of climate change for some major flooding and storm events in 2012.

But that does not mean the question should not be asked. As Bob Ward from Grantham Research Institute on Climate Change and the Environment at LSE told Energydesk: 'The right question is not "was it caused by climate change?" But "what impact has climate change had on it?" He added: 'It would be unlikely an almost one degree increase in global temperature would have no part to play in extreme weather events.'

What Ward and a group of scientists around the world seem to be saying is that the weather events this winter fit into weather patterns and trends that are consistent with the basic physics of climate change.

For instance, the WMO said: 'We do expect to see an increase in extreme heat and precipitation. Already dry regions are expected to become drier and wet ones wetter under IPCC scenarios.'

Energydesk took a look at the most recent evidence on the links between climate change and the types of events we've experienced this winter and attempted to unpick how strong these are.

1. Australia's heatwaves

What's happening?

Last year was Australia's hottest on record with temperatures 1.2 degrees above the average, a year in which Australia's Bureau of Meteorology had to add a new colour to its maps to represent new extremes in heat.

In December 2013, extreme heat developed over southern, central and eastern Australia, with especially high temperatures in the Australian interior into the new year, according to the bureau.

What does this mean for people?

A litany of issues: 100 blazes burning in South Australia, Victoria and New South Wales; New South Wales and Queensland are suffering drought; heart attacks surged 300% during the heatwave; a spike in deaths in Victoria; blackouts; the Australian Open had to stop play after tennis players collapsed.

What the scientists say about links between heatwaves and climate change?

Essentially, the climate in Australia has warmed by about a degree since 1950, and the off-the-charts heatwaves of 2013 is in line with this trend say government meterologists. Average temperatures are projected to be one to five degrees C more by 2070.

The link between climate change and record heatwaves is clear, according to Australian scientists at the Climate Council – they created the body after the Australian Climate Commission was unceremoniously axed by PM Tony Abbott's government.

The council also says climate change is increasing the risk of bushfires.

Scientist says...

Peter Stott, head of climate monitoring at the UK Meteorological Office Hadley Centre told the *Financial Times*: 'Rising temperatures in Australia are a signal of climate change that has emerged very clearly from recent analysis. Last year's temperatures were a long way outside the envelope of variability that we would expect in the absence of climate change.'

Where else in the world?

Argentina has also witnessed one of the worst heatwaves on record at the end of December, according to the WMO. Extreme warmth settled over Russia towards the end of 2013, according to the National Oceanic and Atmosphere Administration.

2. UK flooding and storms

What's happening?

The UK has been battered by one storm after another since the start of December, with a series of storms tracking in off the Atlantic bringing strong winds and heavy rain.

According to the MET office December and January's rainfall was 'one of, if not the most, exceptional periods of winter rainfall in the last 248 years'.

PM David Cameron not only put his wellies on for a photo op but said he 'strongly suspects' the climate change is causing more 'abnormal weather events' such as the floods the UK has been seeing this winter. Even Princes William and Harry have been lugging sandbags.

The British Geological Survey warned that floods could last for months in some areas.

What does this mean for people?

Nearly 6,000 homes flooded across the UK, tens of thousands without electricity, rail networks disrupted, cost of the clean-up could be £1 billion to £3 billion depending on duration.

What does the science say about the link with the flooding and climate change?

What we can say is that it doesn't look like a coincidence that four of the five wettest years recorded in the UK have happened since 2000 at the same time as have also the seven warmest years. As the Met Office pointed out in its recent report, there is an increasing evidence showing heavy rainfall is becoming heavier. They say this is consistent with what you'd expect from basic physics; the atmosphere in a warmer world holds more water vapour = more intense downpours.

The Met Office also linked the UK's storminess with an erratic jet stream – the belt of strong winds circling the planet – over the Pacific Ocean and North America. The North Atlantic jet stream, which blows in storms from the west towards the UK, has been 30% stronger than normal, which links to exceptional wind patterns in the stratosphere with a very intense polar vortex – which has also been affecting weather in the US and Canada.

This whole process was driven by higher than normal ocean temperatures in the West Pacific that was most probably linked to climate change.

Scientist says...

Professor Myles Allen, University of Oxford, said: 'There are simple physical reasons, supported by computer modelling of similar events back in the 2000s, to suspect that human-induced warming of the climate system has increased the risk of the kind of heavy rainfall events that are playing a major role in these floods.'

Where else in the world?

In December Brazil saw torrential rains that saw at least 22 people killed and tens of thousands made homeless.

3. US and Canada cold snap

What's happening?

Extremely cold weather in the US and Canada over the past couple of months – even the Niagara Falls froze. The pattern established in December has continued till mid-February, says the WMO.

In an unprecedented move, the White House released a video explaining how the freezing spell was linked to climate change. The White House also organised a Google+ Hangout on the polar vortex phenomenon.

What does it mean for people?

The freezing weather left around 21 people dead, cancelled 4,000 flights in one day causing traffic chaos across the country and costing the economy an estimated $5 billion.

What does the science say about the link with the cold snap and climate change?

It might seem counter-intuitive, but the cold snap is likely to be linked to climate change.

A split between the Pacific and Atlantic jet streams – with its root in warming ocean temperatures in the Pacific – has resulted in colder air being carried south over North America.

Warming sea ice has also been involved in the frigid weather according to recent studies. It has been implicated in a huge meander in the jet stream over North America resulting in warm weather over Alaska and the west of the US while the rest of the US and Canada freezes.

Another theory about melting Arctic ice driving weather changes is the 'Arctic Paradox' or 'Warm Arctic – Cold Continent' pattern.

Research suggests that as more Arctic sea ice is melting in the summer, the Arctic Ocean warms and radiates heat into the back into the atmosphere in winter.

This disturbs the Polar Vortex – essentially a pattern of strong winds circulating around a low-pressure system that normally sits over the Arctic in the winter – bringing relatively mild conditions to the Arctic while places far to the south bear the brunt of freezing winds. The Polar Vortex is stronger than normal, with increased winds around the vortex, and the vortex has distorted and its core has extended down over Canada.

Scientist says...

In the White House video, President Obama's science and technology advisor, Dr John Holdren, said: 'A growing body of evidence suggests that the kind of extreme cold experienced by the United States is a pattern we can expect to see with increasing frequency as global warming continues.

'I believe the odds are that we can expect as a result of global warming to see more of this pattern of extreme cold in the mid-latitudes and some extreme warm in the far north.'

4. Super-typhoon Haiyan

What happened?

Hitting the Philippines with winds of 310km/h, typhoon Haiyan was the strongest tropical cyclone to make landfall in recorded history. The devastation in coastal areas such as Tacloban was principally caused by a six metre storm surge that carried away even concrete buildings.

What did it mean for people?

Haiyan killed 5,000 people, flattened islands and damaged more than a million houses. Humanitarian disaster.

What does the science say about the link with super-typhoon and climate change?

There's a lot of discussion among scientists over whether storms are getting worse in a warming world. Meteorologists say it's impossible to blame climate change for individual storms.

Scientists think it's plausible that tropical storm activity will rise as the planet warms, on balance. This is despite the effects of increasingly strong 'shearing' winds to prevent the formation of storms or dissipate them.

There is some evidence linking climate change to increasing storm intensity over the past three decades, especially in the north Atlantic where data is available, as described in the IPCC 5 report. But in other places, such as the Northwest Pacific basin, our knowledge is more sketchy because of a lack of data.

One factor in the destruction Haiyan reaped that could be linked to climate change is the fact that rising sea levels, caused by global warming, contributed to the volume of water of the storm surge.

Scientist says...

Julian Heming, a tropical storm prediction scientist at the Met Office told *The Telegraph*: 'We need to look at long-term climate models before we can be certain. But the indications are that the frequency of the storms may decrease – but their intensity will increase.'

Where else in the world?

Climate change was linked to Hurricane Sandy, which hit the east coast of the US in October 2013. The storm was exacerbated by a 'blocking high' of cold air coming down from Canada, a phenomenon linked to global warming by climate scientists.

14 February 2014

⇨ The above information is reprinted with kind permission from Energydesk. Please visit http://energydesk.greenpeace.org for further information.

Climate change and nature

Climate change is already affecting nature in many parts of the world. Rising temperatures affect plants and animals directly, and climate impacts such as changes in rainfall patterns and ocean currents also disrupt the natural world. Ocean acidification is a growing threat to marine life.

Climate change adds to other factors affecting plants and animals, such as loss of habitat, competition from invasive species and disease. Altogether, species are going extinct at the fastest rate known in Earth history.

Global trends

Species are going extinct at a rate that may be 1,000 times faster than the natural rate, leading to warnings that we are experiencing the 'Sixth great extinction' in Earth history.

Climate change is one among many factors driving this trend. Others include habitat loss, wildlife trade, pollution and over-extraction. A recent assessment of 16,857 species by the International Union for the Conservation of Nature found that up to 9% of all birds, 15% of amphibians and 9% of corals are both highly vulnerable to climate change and already threatened with extinction.

The same study found that 24-50% of birds, 22–44% of amphibians and 15–32% of corals are highly vulnerable to warming of 2°C above pre-industrial levels. The global average temperature has already risen by about 0.8°C. If greenhouse gas emissions continue rising at the current rate, the UN Intergovernmental Panel on Climate Change forecasts warming of 2.6-4.8°C by 2100. A recent study found that if emissions continue rising at their record-breaking rate, one in six of the world's species faces extinction. These are global average figures; in some parts of the world, change is occurring much faster.

The climate is also forecast to become more variable in some parts of the world, with more frequent extreme events such as heatwaves and floods. These put additional stress on plants and animals.

Life on the move

Climate change is already causing many species to shift their geographical ranges. In general, they are moving away from the Equator towards the Poles, and from lower to higher ground, as temperatures rise.

For example, the British Comma butterfly has moved 137 miles northward in the past two decades, while geometrid moths on Mount Kinabalu in Borneo have shifted uphill by 59 metres in 42 years. Dartford warblers have been steadily moving northwards in the UK, while on the southern edge of their range, in Spain, they are declining.

Speed counts – both how fast the climate changes, and the species' capacity to move in response. Rapid warming this century is likely to exceed the maximum speed at which many species, in many situations, can disperse or migrate.

Species occupying extensive flat landscapes are particularly vulnerable because they must disperse over longer distances to keep pace with shifting climates.

Those that are usually stationary, such as plants and corals, are especially vulnerable. For example, the Quiver tree of Southern Africa is unable to grow and disperse seeds quickly enough to keep up with a fast-changing climate.

Problems can result if species from the same natural system moving at different rates, UN research predicts that vegetation in mountain gorilla habitat will be dramatically altered by climate change. Warming will push certain plant species further up the mountains, leaving the gorillas with fewer food sources.

Marine life

Climate change and acidification are altering ocean ecosystems in profound ways. Water temperatures are rising, the ocean is slowly turning more acidic as it absorbs carbon dioxide from the air, and areas of the ocean are becoming depleted of oxygen (hypoxia).

In many regions, fisheries will be severely affected. Fish stocks will move – this is already happening in the North Sea and other areas around the UK coasts, for example.

Impacts of climate change and acidification are exacerbated by other factors such as overfishing, habitat loss and pollution. Coral reef ecosystems are declining rapidly, with the risk of potential collapse of some coastal fisheries.

Bees, trees and carbon

Climate change is affecting pollinators such as bees, which provide a vital ecosystem service in maintaining wild plants and many food crops. The UN Intergovernmental Panel on Climate Change concludes that 'after land-use changes, climate change is regarded as the second most relevant factor responsible for the decline of pollinators'.

Forests in temperate regions such as Europe and North America are showing signs of climate stress, with changes in the frequency of fires, insect damage and outbreak of disease. Heatwaves, drought and other extreme events also have an impact.

The extensive forest fires in Russia during the exceptionally hot and dry summer of 2010 are one example. Forests and trees play a critical role in the water cycle, absorb CO_2 from the air, and protect communities against coastal storms.

Our natural world is affected by climate change in many ways, but conserving and managing biodiversity is critical to address climate change. Soils, forests and oceans hold vast stores of carbon, and the way managed habitats are used will affect how much of that carbon is released into the atmosphere.

Europe and the UK

In Europe, by 2080 approximately 60% of plants and vertebrates (animals with a backbone or spinal column) will no longer have favourable climates within protected areas.

The UK has a broadly oceanic climate, with warm summers, mild winters and relatively high rainfall. According to the government, management of forests will need to adapt by mixing tree species and changing the timing of operations to address increased risks from pests, disease, wind and fire.

1 May 2015

⇨ The above information is reprinted with kind permission from the Energy and Climate Intelligence Unit. Please visit www.eciu.net for further information.

Warming oceans speeding up climate change cycle

Scientists have identified another consequence of global warming that is likely to accelerate climate change still further.

By Tim Radford

The warming oceans could start to return more carbon dioxide to the atmosphere as the planet warms, according to new research.

And since 70% of the planet is covered by clear blue water, anything that reduces the oceans' capacity to soak up and sequester carbon could only make climate change more certain and more swift.

It is a process that engineers call 'positive feedback'. And under such a cycle of feedback, the world will continue to get even warmer, accelerating the process yet again.

Many such studies are, in essence, computer simulations. But Chris Marsay – a marine biochemist at the UK's National Oceanography Centre in Southampton – and colleagues based their results on experiments at sea.

Sediment traps

They report in the *Proceedings of the National Academy of Sciences* that they examined sediment traps in the North Atlantic to work out what happens to organic carbon – the tissue of the living things that exploit photosynthesis, directly or indirectly, to convert carbon dioxide – as it sinks to the depths.

Sooner or later, much of this stuff gets released into the sea water as carbon dioxide. This is sometimes called the ocean's biological carbon pump. In deep, cold waters, the process is slow. In warmer, shallower waters, it accelerates.

And as there is evidence that the ocean is responding to atmospheric changes in temperature, both at the surface and at depth, the study suggests that 'predicted future increases in ocean temperatures will result in reduced CO_2 storage by the oceans'.

The research was conducted on a small scale, in a limited stretch of ocean, so the conclusion is still provisional – and, like all good science, will be confirmed by replication. But it is yet another instance of the self-sustaining momentum of global warming.

Such positive feedbacks are already at work in high latitudes. Ice reflects sunlight, and therefore the sun's heat. So as the Arctic ice sheet steadily diminishes over the decades, more and more blue water is available to absorb heat – and accelerate warming.

The same gradual warming has started to release another greenhouse gas trapped at the ocean's edge. Natural 'marsh gas', or methane, is stored in huge masses, 'frozen' as methane hydrate in cold continental shelves.

Methane exists in much smaller quantities than carbon dioxide, and has a shorter life in the atmosphere, but is far more potent, volume for volume, as a greenhouse gas.

Researchers at the Arctic University of Norway in Tromso reported last month in *Geophysical Research Letters* that once-frozen methane gas was leaking from thawing ocean floor off Siberia.

Some of this thaw is natural, and perhaps inevitable. But some is connected with human influence and could accelerate.

Alexei Portnov, a geophysicist at the university's Centre for Arctic Gas Hydrate, Climate and Environment says: 'If the temperature of the oceans increases by two degrees, as suggested by some reports, it will accelerate the thawing to the extreme. A warming climate could lead to an explosive gas release from the shallow areas.'

Biological origin

Arctic methane, like ocean organic carbon, has a biological origin. It is released by decaying vegetation under marshy conditions and tends to form as a kind of ice at low temperatures and high pressures, much of it along continental shelves that, at the height of the Ice Ages, were above sea level.

The International Union for Conservation of Nature also reminded the world last month that the ocean plays a vital role in climate, and that plankton, fish and crustaceans could be considered as 'mobile carbon units'.

In this sense, the fish in the sea are not just suppers waiting to be caught, but are important parts of the planetary climate system. The healthier the oceans, and the richer they are in living things, the more effective they become at soaking up atmospheric carbon.

'The world is at a crossroads in terms of climate health and climate change,' said Dan Laffoley, Vice-Chairman of the IUCN World Commission on Protected Areas, introducing a new report on the marine role in the carbon cycle.

'Neglect the ocean and wonder why our actions are not effective, or manage and restore the ocean to boost food security and reduce the impact of climate change. The choice should be an easy one.'

This article was produced by the Climate News Network

9 January 2015

⇨ The above information is reprinted with kind permission from Responding to Climate Change (RTCC). Please visit www.rtcc.org for further information.

Moving stories

The voices of people who move in the context of environmental change.

By Alex Randall

Pakistan

'The rains came in the middle of the night, while most people were sleeping. When we woke up, there was water of about two to three feet and we did not know how to escape, because our village is far from the main road. The water was very dirty because the floods had damaged our [sanitation and water] facilities. I was very pregnant at the time, and our livestock are our livelihood so we didn't want to leave them to die, so we did not know what to do. We were rescued in boats by the army and non-governmental organisations (NGOs). We are thankful to be alive, but we lost our livestock and now we are trying to rebuild our livelihood by starting from the beginning.' – Fatay and Zulaikar, husband and wife of a pastoralist family in Badin district[1].

'I go to get registered [as an IDP (internally displaced person)] and they dismiss me. I don't want to live here. I don't want my children out on the streets. In my village I have little but I look after my family. They throw food at me like I am a beggar. I have never begged for anything in my life, why do they treat me like this?' – Shauquat Ali, displaced tenant farmer and father of nine[2].

'The water came at night and we didn't have time to save our belongings; we had to choose whether to save our children and ourselves or our property and assets, so we chose to save our kids. We left everything and ran to save our lives.' – Unnamed survivor of the 2010 floods[3].

1 Climate and Development Knowledge Network (CDKN). 2012. *Disaster-proofing your village before the floods – the case of Sindh, Pakistan.* [ONLINE] Available at: http://bit.ly/MY47au. [Accessed 21 August 13]

2 Aljazeera. 2010. *Pakistan flood victims need dignity as much as aid.* [ONLINE] Available at: http://blogs.aljazeera.com/blog/asia/pakistan-flood-victims-needdignity-much-aid. [Accessed 21 August 13].

3 World Food Programme. 2012. *Pakistan Flood Victims 'Left With Nothing'.* [ONLINE] Available at: http://bit.ly/16CUBo0. [Accessed 21 August 13].

Context

Pakistan is highly exposed to the impacts of climate change. The IPCC have associated increasing temperatures with the severity[4] of the monsoon rains and predict an increase in severity. Northern Pakistan faces increased risk of flooding and landslides[5]. An increase of cyclonic activity will impact Southern Pakistan[5] and the city of Karachi is at high risk from sea-level rise, prolonged cyclonic activity, and greater salt-water intrusion[5]. Pakistan's vulnerability is increased due to its reliance upon water from the Indus river and tributaries, which supply two thirds of the water the country uses for irrigation and domestic use[6]. The Indus is fed by the Himalayan glaciers, which are receding significantly, with the likelihood of them disappearing by the year 2035[7]. Other factors make Pakistan vulnerable to the impacts of climate change. Nearly half of the population is dependent on agricultural livelihoods; there is considerable rural poverty, urban unrest, land degradation and shortfalls in food

4 M.L. Parry, O.F. Canziani, J.P. Palutikof, P.J. van der Linden and C.E. Hanson (eds) (2007). *Contribution of Working Group II to the Fourth Assessment Report of the Intergovernmental Panel on Climate Change, 2007.* Cambridge, United Kingdom and New York, NY, USA: Cambridge University Press.

5 Asian Development Bank, 2012. *Addressing Climate Change and Migration in Asia and the Pacific.* 1st ed. Manila: ADB.

6 Pangare, G., Das, B., Lincklaen Arriens, W., and Makin, I. (2012). *WaterWealth? Investing in Basin Management in Asia and the Pacific.* New Delhi, India: Academic Foundation.

7 M.L. Parry, O.F. Canziani, J.P. Palutikof, P.J. van der Linden and C.E. Hanson (eds) (2007). *Contribution of Working Group II to the Fourth Assessment Report of the Intergovernmental Panel on Climate Change, 2007.* Cambridge, United Kingdom and New York, NY, USA: Cambridge University Press.

production[8]. Further urbanisation and industrialisation place more pressure upon water supplies which are already threatened by climate change.

In July 2010, Pakistan was affected by heavy monsoon rains, which led to massive flooding in the Indus River basin. More than ten million people were displaced, with about 20% of the country under water. The death toll was around 2,000. The provision of international aid was widely considered insufficient, with millions of farmers housed in refugee camps, and crops and cattle destroyed[9]. Flooding struck again in 2011. The disaster affected 18 million people and destroyed 1.7 million homes[10].

In August 2012, following the monsoon floods, the region of Tharparkar experienced significant drought, forcing 600,000 people dependent on rainfed agriculture to internally migrate[11]. The International Federation of Red Cross (IFRC) put the number of displaced people at 250,000. The total number of people affected stands at 4.4 million[12].

Floods and natural disasters cause considerable forced migration within Pakistan. Pakistan receives roughly 8% of the total global funding available for dealing with displacement. There are 745,000 IDPs, the majority fleeing from fighting in Federally Administered Tribal Areas (FATA)[12]. But despite such a large IDP population, Pakistan is also a destination for international migrants in the region. It is the top destination for Somali refugees[13]. There are currently 1.6 million Afghan refugees in Pakistan. The country is currently experiencing the world's largest protracted refugee situation[14].

As well as vast numbers of displacees and the absorption of neighbouring refugees, Pakistan also has a long history of voluntary migration, which is largely split between unskilled labourers travelling to the Emirates and Dubai and more skilled workers heading for Europe and the US. Pakistani diasporas are amongst the largest and most extensive in the world, supplying remittances to families in Pakistan of $12 billion a year[15]. The IPCC predicted that 'circular migration patterns, such as those punctuated by shocks of migrants following extreme weather events, could be expected'. This is supported by the Asia Development Bank which suggests that 'environmental factors are already an important driver in migration' and that 'floods, cyclones and desertification have led in recent years to significant population movements, mostly from rural to urban areas'.[15]

The Arctic

'About five years ago the sea ice used to take longer to melt. It lasted about ten months but now it's only eight months. This harms our way of life, our way of hunting, our way of fishing, and our way of travelling

8 M.L. Parry, O.F. Canziani, J.P. Palutikof, P.J. van der Linden and C.E. Hanson (eds) (2007). *Contribution of Working Group II to the Fourth Assessment Report of the Intergovernmental Panel on Climate Change, 2007.* Cambridge, United Kingdom and New York, NY, USA: Cambridge University Press.

9 Asian Development Bank, 2012. *Addressing Climate Change and Migration in Asia and the Pacific.* 1st ed. Manila: ADB.

10 Office for the Coordination of Humanitarian Affairs, 2011. *Pakistan Media Factsheet.* [ONLINE] Available at: http://bit.ly/151MVsY. [Accessed 21 August 13].

11 Shaikh, S; Tunio, S. 2012. *Extreme weather in Pakistan pulls many into downward spiral.* Thomson Reuters Foundation, 1 October.

12 International Organisation for Migration, 2012. IOM *Pakistan response - six months on.* 1st ed. Islamabad: IOM.w

13 Zetter, R (Eds), 2012. *World Disasters Report. Focus on Forced Migration and Displacement.* 1st ed. Geneva: International Federation of Red Cross and Red Crescent Societies.

14 United Nations High Commission for Refugees. 2013. *2013 UNHCR country operations profile - Pakistan.* [ONLINE] Available at: http://bit.ly/10SCcR6. [Accessed 21 August 13].

15 Asian Development Bank, 2012. *Addressing Climate Change and Migration in Asia and the Pacific.* 1st ed. Manila: ADB.

from one place to another.' – Charlie Nakqashuk, Pangnirtung, Nunavut[16].

'If we relocate during the summer, we'd need a lot of barges to move everything… The thought of moving our village is very sad because Shishmaref is the place where I grew up. Shishmaref is a great place to live because everyone knows each other. If we move, it would probably bring our community closer together. It could be totally the opposite though because some people might just move to a new place. If we move, it won't be the same because it wouldn't be the Shishmaref that everyone knows.' – Allison 'Anisaaluk' Nayokpuk, Shishmaref, Alaska.[16]

Context

The Arctic's indigenous people have traditionally led nomadic lives. However, over the last half century there has been a general decrease in the mobility of indigenous Arctic people. Historically the primary cause of this has been the desire of central governments to provide services such as health and education, which to some extent required nomadic people to settle into small towns and villages. This sedentarisation has not always been voluntary and there are examples of forced settlement from across the Arctic regions[17], especially in the Russian far north[18]. Another major trend has been the outmigration of young people in search of work and education in other locations.

A number of significant climatic changes have taken place in the Arctic: air temperatures over extensive land areas have increased, sea ice has thinned and declined, Atlantic water flowing into the Arctic has warmed, and terrestrial permafrost and Eurasian spring snow has decreased[19]. These environmental changes have a global impact as less solar radiation is reflected away from the surface of the Earth causing more heat to be absorbed and more snow and ice to melt, leading to rising sea levels[20]. Changes to freshwater and marine systems and fisheries are already being experienced within the region, particularly by indigenous Arctic communities.

Climate change has become an additional emerging force impacting on the mobility of indigenous Arctic peoples in a number of ways. Changes in habitat are altering the way in which people forage and hunt. Altered weather patterns are changing the usual animal herding patterns, as people are forced to move with their herds in order to continue hunting[21]. Traditional hunting routes are also changing or disappearing as thinner winter ice means crossing rivers and lakes becomes harder or impossible. Many indigenous Arctic communities depend on barges for supplies, but increased numbers of storms and the melting of a layer of protective sea ice has caused many barge landing sites to become unusable due to erosion. In a number of cases this has forced people to find new barge sites. However, in one notable case the disappearance of barge sites and other coastal erosion resulted in the village of Newtok facing relocation.

Climate change may also result in new industry and activity in Arctic regions, which will have implications for the movement of indigenous people. As ice thins and melts, new opportunities for fossil fuel extraction and mining emerge. This could mean that Arctic people move in order to find work in these new operations, and that people from other areas will increasingly move into the Arctic. However, it is unclear what impact these new operations will have on the culture, health and well-being of indigenous people: 'Any economic advantage that might trickle down to the Inuit cannot compensate for the hugely negative effects of climate change on their health and well-being.'[22] Further, the fact that the melting Arctic has created an opportunity for more fossil fuel extraction is deeply and tragically ironic.

References

To see full references please visit: http://climatemigration.org.uk/wp-content/uploads/2014/01/MovingStories.pdf

⇨ The above information is reprinted with kind permission from the Climate Outreach and Information Network. Please visit www.climateoutreach.org.uk or www.climatemigration.org.uk for further information.

16 Kelman, I. (2012). *Many Strong Voices. Available: Many Strong Voices.* [Accessed 25 July 2013].

17 Backgrounder - Apology for Inuit High Arctic Relocation. *Aboriginal Affairs and Northern Development Canada.* http://www.aadnc-aandc.gc.ca/eng/1100100015426/1100100015427

18 Nuykina, Elena. *Resettlement from the Russian North: an analysis of state-induced relocation policy.* Arctic Centre Reports 55.

19 IPCC, 2007: Climate Change 2007: Cambridge University Press, Cambridge, United Kingdom and New York, NY, USA. Contributions of Working Group II, Climate Change 2007, Impacts, Adaption and Vulnerability, Chapter 15, 15.1.1, p 656.

20 IPCC, 2007: Climate Change 2007: Cambridge University Press, Cambridge, United Kingdom and New York, NY, USA. Page 656, Chapter 15, Polar Regions, Arctic and Antarctic.

21 Ferris, B (2013). *A Complex Constellation: Displacement, Climate Change and Arctic Peoples:* Brookings-LSE Project on Internal Displacement.

22 Indigenous peoples global summit on climate change, 2009.

Chapter 2 The climate debate

How does the IPCC know climate change is happening?

***An article from* The Conversation.**

THE CONVERSATION

By Mark Maslin, Professor of Climatology at UCL

Climate change is one of the few scientific theories that makes us examine the whole basis of modern society. It is a challenge that has politicians arguing, sets nations against each other, queries individual lifestyle choices, and ultimately asks questions about humanity's relationship with the rest of the planet.

The Intergovernmental Panel on Climate Change published its synthesis report on 2 November, a document that brings together the findings from the IPCC's three main working groups. It reiterates that the evidence for climate change is unequivocal, with evidence for a significant rise in global temperatures and sea level over the last 100 years. It also stresses that we control the future and the magnitude of shifting weather patterns and more extreme climate events depends on how much greenhouse gas we emit.

This is not the end of the world as envisaged by many environmentalists in the late 1980s and early 1990s, but it will mean substantial, even catastrophic challenges for billions of people.

Causes of climate change

Greenhouse gases absorb and re-emit some of the heat radiation given off by the Earth's surface and warm the lower atmosphere. The most important greenhouse gas is water vapour, followed by carbon dioxide and methane, and without their warming presence in the atmosphere the Earth's average surface temperature would be approximately -20°C.

While many of these gases occur naturally in the atmosphere, humans are responsible for increasing their concentration through burning fossil fuels, deforestation and other land use changes.

Although carbon dioxide is released naturally by volcanoes, ecosystems and some parts of the oceans, this release is more than compensated for through the carbon absorbed by plants and in other ocean regions, such as the North Atlantic. Had these natural carbon sinks not existed, CO_2 would have built up twice as fast as it has done. Records of air bubbles in ancient ice show us that carbon dioxide and other greenhouse gases are now at their highest concentrations for more than 800,000 years.

Evidence for climate change

The IPCC presents six main lines of evidence for climate change.

We have tracked the unprecedented recent rise in atmospheric carbon dioxide and other greenhouse gases since the beginning of the Industrial Revolution.

We know from laboratory and atmospheric measurements that greenhouse gases do indeed absorb heat when they are present in the atmosphere.

We have tracked significant increase in global temperatures of 0.85°C and sea level rise of 20cm over the past century.

We have analysed the effects of natural events such as sunspots and volcanic eruptions on the climate, and though these are essential to understand the pattern of temperature changes over the past 150 years, they cannot explain the overall warming trend.

We have observed significant changes in the Earth's climate system including reduced snowfall in the Northern Hemisphere, retreat of sea ice in the Arctic, retreating glaciers on all continents, and shrinking of the area covered by permafrost and the increasing depth of its active layer. All of which are consistent with a warming global climate.

We continually track global weather and have seen significant shifts in weather patterns and an increase in extreme events. Patterns of precipitation (rainfall and snowfall) have changed, with parts of North and South America, Europe and northern and central Asia becoming wetter, while the Sahel region of central Africa, southern Africa, the Mediterranean and southern Asia have become drier. Intense rainfall has become more frequent, along with major flooding. We're also seeing more heatwaves. According to the US National Oceanic and Atmospheric Administration (NOAA) between 1880 and the beginning of 2014, the 13 warmest years on record have all occurred within the past 16 years.

Future changes

The continued burning of fossil fuels will inevitably lead to further climate warming. The complexity of the climate system is such that the extent of such warming is difficult to predict, particularly as the largest unknown is how much greenhouse gas we will emit over the next 85 years.

The IPCC has developed a range of emissions scenarios or Representative Concentration Pathways (RCPs) to examine the possible range of future climate change. Using scenarios ranging from business-as-usual to strong longer-term decline in emissions, the climate model projections suggest the global mean surface temperature could rise by between 2.8°C and 5.4°C by the end of the 21st century.

The sea level is projected to rise by between 52cm and 98cm by 2100, threatening coastal cities, low-lying deltas and small islands. Snow cover and sea ice are projected to continue to reduce, and some models suggest that the Arctic could be ice-free in late summer by the latter part of the 21st century. Heat waves, extreme rain and flash flood risks are projected to increase, threatening ecosystems and human settlements, health and security.

These changes will not be spread uniformly around the world. Faster warming is expected near the poles, as the melting snow and sea ice exposes the darker underlying land and ocean surfaces which then absorb more of the Sun's radiation instead of reflecting it back to space in the way that brighter ice and snow do. Indeed, such 'polar amplification' of global warming is already happening.

Changes in precipitation are also expected to vary from place to place. In the high-latitude regions (central and northern regions of Europe, Asia and North America) the year-round average precipitation is projected to increase, while in most sub-tropical land regions it is projected to decrease by as much as 20%, increasing the risk of drought.

In many other parts of the world, species and ecosystems may experience climatic conditions at the limits of their optimal or tolerable ranges or beyond. Human land use conversion for food, fuel, fibre and fodder, combined with targeted hunting and harvesting, has resulted in species extinctions some 100 to 1,000 times higher than background rates. Climate change will only speed things up.

Solutions

The IPCC synthesis set in stark terms the global challenge of reducing greenhouse gas emissions. To keep global temperature rise below 2°C then global carbon emission must peak in the next ten years and from 2070 onward must be negative: we must start sucking out carbon dioxide from the atmosphere.

Despite 30 years of climate change negotiations there has been no deviation in greenhouse gas emissions from the business-as-usual pathway so many feel keeping the climate change to less than 2°C will prove impossible.

The failure of the international climate negotiation, most notably at Copenhagen in 2009, set back meaningful global cuts in emissions by at least a decade. Anticipation is building for the Paris conference in 2015 and there are some glimmers of hope.

China, now the largest greenhouse gas polluter in the world, has discussed instigating a regional carbon-trading scheme which if successful would be rolled out across the whole country. Meanwhile the US, which has emitted a third of all the carbon pollution in the atmosphere, has placed the responsibility for regulating carbon dioxide emissions under the Environment Protection Agency, away from political wrangling in Washington.

Support and money are also needed to help developing countries mitigate carbon emissions and adapt to inevitable climate change. Trillions of dollars will be invested in energy over the next 15 years to keep pace with increasing demand – what we must do is ensure that it is directed towards developing cheap, clean, secure energy production rather than exploiting fossil fuels. We must also prepare for the worst and adapt. If implemented now, much of the costs and damage that could be caused by changing climate can be mitigated.

Climate change challenges the very way we organise our society. It needs to be seen within the context of the other great challenges of the 21st century: global poverty, population growth, environmental degradation and global security. To meet these challenges we must change some of the basic rules of our society to allow us to adopt a much more global and long-term approach and in doing so develop a solution that can benefit everyone.

3 November 2014

⇨ The above information is reprinted with kind permission from The Conversation Trust (UK). Please visit www.theconversation.com for further information.

Who are the climate change deniers?

Republicans are far more sceptical of human-caused climate change than most Americans, who also tend to think that climate change will have a noticeable impact within their lifetimes.

By Kathy Frankovic

Nine per cent of Americans in the latest *Economist*/YouGov Poll say the earth's climate is not changing. 24% say there won't be any climate change effects, like rising sea levels and more intense storms, in their own lifetimes. These climate change deniers are better-off than average, more conservative, and more Republican.

20% of Republicans say the Earth's climate is not changing. Those Republicans who admit it is are nearly twice as likely to say it has nothing to do with human activity as to say it does. For the population as a whole, more than half believe in climate change and attribute it to human activity.

Most climate change deniers don't think there is a scientific consensus that climate change is occurring and is caused by human activity, and nearly half have no trust that whatever climate scientists say would be the truth. That is very different from the opinions of the public overall.

Deniers have little trust in any source when it comes to telling the truth about climate change, but they are more willing to trust military officials to tell the truth. Only a third of deniers have no trust in the military. Although their trust in Republican politicians is higher than their trust in Democrats or government officials, it is not much better than their trust in climate scientists.

Nearly half of Republicans (45%) don't expect they will have to deal with the effects of climate change in their lifetimes, claiming they won't see rising sea levels, more severe storms and other problems in their lifetimes.

Republicans are half as likely as the public overall to say they have already seen climate change. 44% of all Americans say they have, compared with only 22% of Republicans. The difference is partly due to geography: those who live away from the two coasts are less likely to report climate change activity. And those living in the Northeast and the West are less likely to be Republicans.

But for most Americans, climate change is real. It is also something that developed countries have a role in solving. 60% believe the most developed countries have a greater responsibility than other countries to control greenhouse gases. Most agree that the United States has a particularly important role to play, even at the risk of becoming less competitive. More than one in four aren't sure.

In addition, a majority thinks the United States has a responsibility to help poorer countries with the effects of climate change.

As on many other climate change questions, Republicans take the opposing view.

There is some hope that government action can slow global warming. By 47% to 30%. Americans think governments can act now to slow climate change.

Americans divide when it comes to evaluating Barack Obama's performance on handling climate change – and the environment overall. Just about as many disapprove of how he is handling them as approve.

Those numbers are much better that the assessments of the President's handling of several foreign policy concerns.

13 August 2014

⇨ The above information is reprinted with kind permission from YouGov. Please visit www.yougov.co.uk for further information.

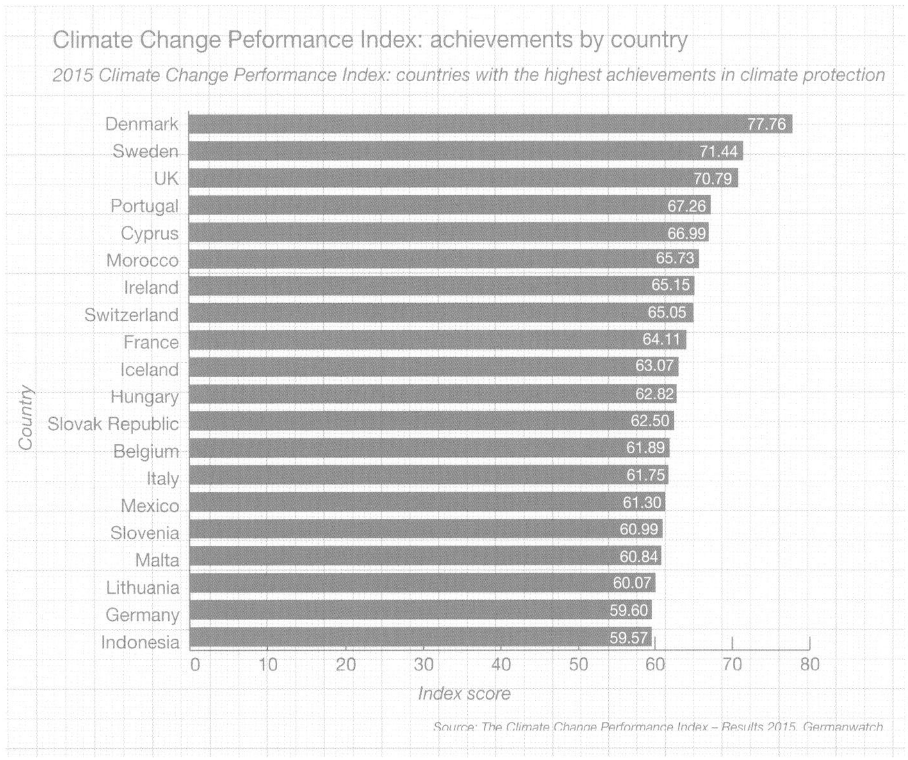

The real 'deniers' in the climate change debate are the warmists

Those who believe that the world faces a catastrophe from global warming dismiss anyone who dares question their beliefs.

By Christopher Booker

Among the more useful things psychologists have come up with in helping us to understand the world is what they call 'projection' – that tendency in people who are not seeing the world straight to 'project' on to others the very failings they suffer from themselves.

One prominent example of this has been the contemptuous itch of those who passionately believe that the world faces a catastrophe from global warming to dismiss anyone daring to question their beliefs as a 'denier'. In fact, it has long been obvious that the only real 'deniers' on the climate front are those true believers themselves, who cannot face up to all the evidence which makes their scare story ever less convincing.

Last week, under the heading 'How Arctic ice has made fools of all those poor warmists', I again wrote about how the state of Arctic ice and those iconic polar bears have long been made the supreme poster-child for their cause. Again and again they have warned us that, within only a few years (or even months), the Arctic would be completely 'ice-free', while the poor polar bears are vanishing as fast as the ice. Yet every time, the evidence turns up to show how they are talking through their hats.

Many times when I have been tempted to quote some of that evidence, no one has got more cross than the denizens of that citadel of warmist groupthink, *The Guardian*. When in 2009 I observed that, after a further severe dip in 2007, Arctic ice levels had made a significant recovery, this so enraged George Monbiot that he rushed into print with a piece headed 'How to prove Christopher Booker wrong in 26 seconds'. He claimed that only the briefest look at the evidence would show how I had got it all absurdly wrong.

Within minutes of posting his article, however, the Great Moonbat had to come back with an apology.'Whoops,' he wrote, 'looks like I've boobed. Sorry folks.' One of his readers had noticed that he had totally misread what I wrote, because he had been looking at the wrong graph.

Last week, after I had been prompted again to write about the Arctic by a new Cryosat satellite study showing that in 2013 its ice volume increased by a staggering 33 per cent, this provoked a strident response from another of the newspaper's attack dogs, Dana Nuccitelli.

Only a stupid 'denialist' like me, he suggested, could be foolish enough to suggest that polar bear numbers in the Arctic were actually rising, not falling – or to deny that temperatures in Greenland have not been rising so dangerously fast that, since 1990, they have been hurtling upwards at an incredible 1.1 °C per decade.

But if Mr Nuccitelli wanted some actual evidence, he might consult a graph recently posted on the blog Notalotofpeopleknowthat, headed 'CBS climate liars are hard at work'. The blogger, Paul Homewood, has meticulously plotted the temperature data from two of Greenland's main weather stations, going right back to 1900. Perhaps surprisingly, these show that the temperature trend in the island that contains a tenth of the world's land ice has not risen an iota. It has remained astonishingly stable.

As for those polar bears, I have been following this story for years, ever since I used to speak to Dr Mitch Taylor, a scientist who has been observing and counting polar bears for many decades, and knows far better than any computer model how much their numbers have been rising. One has only to Google 'polar bear numbers increasing' to find several studies that explain how, since the Seventies, their numbers have steadily risen – in some places by as much as 250 per cent.

But such are the inconvenient truths to which those 'deniers' at *The Guardian* will remain determined to shut their eyes until the world freezes over.

1 August 2015

⇨ The above information is reprinted with kind permission from *The Telegraph*. Please visit www.telegraph.co.uk for further information.

Climate change: the science

Frequently asked questions.

The Committee on Climate Change is the independent advisor to Parliament, and the Government, on climate change issues. We are established by legislation passed by Parliament and are funded to provide independent advice consistent with the legislation. As such, we review existing and emerging evidence on a regular basis. The responses to the questions provided below reflect the current available evidence. We will review the responses as new evidence is produced.

1) The climate has changed in the past. Isn't this just another natural cycle?

There are natural cycles, but what we are seeing now is very different. The main cycle of global change over the last million years has been in and out of 'ice ages', in which Earth's average surface temperature changed by about 4–7°C. For the last 10,000 years we have been in a warm phase between ice ages, meaning we would expect gradual cooling rather than the rapid further warming seen since 1950. The pattern of warming also matches what we expect from our emissions, and does not match that from other natural influences alone such as the Sun.

Scientists predict the Earth will warm 1.7–5.4°C by the end of this century without concerted efforts to reduce greenhouse gas emissions. This pace of global change is far faster than any of the ice age cycles.

2) Why trust climate models?

A model is simply a mechanism or tool for using our knowledge to investigate some aspect of the real world. Models are used in many areas of science, especially where controlled experiments on the real system are very difficult (such as astronomy, the behaviour of complex molecules and climate science). Building, testing and disputing models is a fundamental part of science. The inputs to the models are crucial. They are based on a range of evidence, including controlled experiments, historical observations and measurements.

There are many different climate models developed by scientists, from simple one-line equations to complex simulators that run on supercomputers. The most complex models calculate the relevant processes of physics, chemistry and biology as thoroughly as possible, producing realistic-looking weather patterns. They do however have known flaws – they do not currently have enough detail to simulate individual clouds, for example, and their representation of the net effects of clouds is not completely accurate. But scientists are working continually to improve them and they are able to reproduce many of the large-scale climate changes seen this century.

Further, predictions of future climate do not solely rely on these complex models. Evidence also comes from basic theory and much simpler models, and from measurements of past climate change.

3) Hasn't there been a pause in global warming?

Average surface temperature has risen more slowly over the last few years. This 'pause' is still a short period of time for the climate, and it does not have a major impact on long-term projections.

Since 1998 global temperature has been rising at a rate of 0.04°C per decade, lower than the longer-term rate of 0.11°C per decade since 1950. The temperature record is noisy due to natural, unpredictable fluctuations which can mask (or add to) underlying changes. Scientists therefore use longer periods (around 30 years or more) to identify robust climate trends.

The cause of the current slowdown is a topic of research. Scientists have identified several plausible causes, from slightly reduced solar heating and a series of small volcanic eruptions to additional storage of heat in the oceans. But there is very strong evidence that the whole climate system is still heating up, and the global surface temperature rise will resume in the coming decades.

Other important indicators of a changing climate have not paused since 1998: Arctic sea ice is decreasing, global sea level is rising, and the hottest days of the year are becoming hotter.

This information was correct at the time of going to print in August 2015.

⇨ The above information is reprinted with kind permission from the Committee on Climate Change. Please visit www.theccc.org.uk for further information.

Climate change scientists urged to be more open to the public about uncertainties

A new report calls for experts to communicate their research more clearly.

By Tom Bawden, Environment Editor

Climate change scientists must be more honest about the limits of their knowledge and uncertainty around predictions if they are to win the trust of the public, according to a new report.

Scientists are under increasing pressure to communicate their research more clearly, to galvanise politicians into taking decisive action to combat climate change, and to help promote their universities.

They are also keen to make their findings meaningful to a public which feels alienated from much climate change research, which is largely abstract and concerned with developments that often lie decades in the future, said Dr Gregory Hollin, of the University of Nottingham.

But this increases the temptation to gloss over any uncertainties in research – an urge they should resist if they don't want to lose credibility, his report says. And while referencing recent events such as floods and heatwaves can make climate change seem more tangible, they are much less scientifically certain as evidence.

'The most meaningful things are often the least certain things and so that potentially leads to difficulties because scientists are being asked to make their results really meaningful, while being incredibly certain. And there are instances when that leads to real tensions,' Dr Hollin told *The Independent*.

He acknowledged in some cases scientists are keen to play down uncertainties to limit the scope for climate sceptics to magnify their doubts and use it to attack research – but this approach makes the problem worse. The research, which Dr Hollin carried out with his colleague Dr Warren Pearce and is published in the journal *Nature Climate Change*, focused on a key press conference in 2013 when the UN Intergovernmental Panel on Climate Change unveiled its latest report.

The scientists repeatedly emphasised the finding that the decade from 2001 onwards was the hottest on record, to bolster the case that man-made climate change is happening and poses a major threat.

However, when asked about the widely perceived slowdown in global warming since the late 1990s – since shown by research in the US to be a fallacy – a scientist dismissed the question by saying that the period was too short to draw any meaningful conclusions.

'A switch to shorter periods made it harder to dismiss media questions about short-term uncertainties in climate science,' Dr Pearce wrote. 'The fact scientists go on to dismiss journalists' concerns about the pause – when they themselves drew upon a similar short-term example – made their position inconsistent.'

9 June 2015

⇨ The above information is reprinted with kind permission from *The Independent*. Please visit www.independent.co.uk for further information.

Climate is changing – but some believe the threat has been exaggerated

79% of people believe that the world's climate is changing, but 39% believe concerns have been exaggerated by scientists.

The International Panel on Climate Change (IPCC) is preparing its fifth Assessment Report on climate change, due out on Friday. The IPCC report brings together 800 scientists in 85 countries and serves as the basis for UN negotiations on CO_2 emissions. This build-up to this year's report has been overshadowed by allegations of political wrangling as sceptics demand an explanation for the 'standstill' in global temperatures since 1998.

A new YouGov survey has found that a majority of the public feel the climate is changing, but a significant minority feel that concerns have been exaggerated.

In total, 79% of the British public think the world's climate is changing, however, 23% think it is not man-made, compared to 56% that believe it is. 7% believe the world's climate isn't changing and 14% don't know.

Belief in man-made climate change is up 6% from our last poll. In June, when the question was last asked, 49% felt that climate change was a result of human activity, 28% felt it was changing but not as a result of human activity, 7% felt the climate was not changing and 16% weren't sure.

The political row this week centres on whether the effects of carbon dioxide emissions have been exaggerated in previous reports and a significant proportion of British public share this view.

39% believe that concerns over climate change have been exaggerated, although 47% believe the threat is as real as scientists have said.

When it comes to the public response to climate change, with the backdrop of rising fuel bills and a cost of living crisis in the UK, only 22% of the public would be willing to see electricity bills rise if the money was spent on cleaner energy. Two-thirds (67%) would not be willing to see their bills rising.

The IPCC won the Nobel Prize for Peace for its previous Assessment Report in 2007.

23 September 2013

⇨ The above information is reprinted with kind permission from YouGov. Please visit www.yougov.co.uk for further information.

40% of adults worldwide have never heard of climate change

Apparently there's still a bit more work to do to communicate that the world is changing.

By Alison Azaria

According to analysis of global climate change awareness and risk perception published in *Nature Climate Change*, around 40% of adults worldwide have never heard of climate change.

The percentage of people unaware of the global phenomenon rises to more than 65% in developing countries such as Egypt, Bangladesh and India, whereas only 10% of the public is unaware in North America, Europe, and Japan.

'The findings indicate that strategies for securing public engagement in climate issues will vary from country to country because different populations perceive climate-related risks very differently,' so say the researchers at Yale Program on Climate Change Communication (YPCCC) which carried out the study.

'In many African and Asian countries, for example, climate risk is most strongly perceived through noticeable changes in local temperatures. In the US, Latin America, Europe and China, however, understanding that climate change is caused by humans increases the public's perception of climate risks.'

Co-author of the study and director of the YPCCC, Anthony Leiserowitz added that the 'contrast between developed and developing countries was striking'.

The research found that globally, education levels tend to be the single biggest predictor of public awareness of climate change. However, these factors vary between countries. In the US, the key predictors of awareness are civic engagement, communication access and education. In China, it is most closely association with education, proximity to urban areas and household income.

3 August 2015

⇨ The above information is reprinted with kind permission from 2degrees. Please visit www.2degreesnetwork.com for further information.

Britons believe in climate change... but do they care?

Survey shows 88 per cent of public believe climate is changing yet a record low of just 18 per cent are 'very concerned' about it.

By Emily Gosden, Energy Editor

The overwhelming majority of Britons believe in climate change but fewer than one in five is very worried about it, new research has revealed.

Despite warnings from UN scientists, politicians and even Prince Charles that time is running out to avoid catastrophic global warming, the number of people describing themselves as 'very concerned' has more than halved over the past decade.

The proportion of Britons who believe the world's climate is changing now stands at 88 per cent – the highest level since 2005, when it hit 91 per cent.

Belief levels are up significantly from only 72 per cent in early 2013, according to a survey of 1,002 people conducted as part of Government-funded research by the Universities of Cardiff and Nottingham.

Severe flooding that hit the Somerset Levels in winter 2013–14 – which David Cameron linked to climate change – may have influenced opinion, the researchers suggested.

The uptick in belief also covered the period in which Typhoon Haiyan, which was also widely linked to global warming, caused devastation in the Philippines and the UN Intergovernmental Panel on Climate Change issued major reports on global warming.

The vast majority – 84 per cent – also believe that climate change is either partly or mainly man-made.

Yet the number describing themselves as 'very concerned' fell to just 18 per cent, the survey shows. That was down from 44 per cent in 2005.

Overall, those who were either very or fairly concerned were still the majority, at 68 per cent, but much lower than 2005 highs of 82 per cent.

Dr Stuart Capstick, one of the report's authors, said: 'Whilst on a technical or "cognitive" level people may be more willing to accept the reality and human aspect of climate change, this does not necessarily translate into very high levels of concern at a more personal, emotional level.

'It is possible also that people have perhaps grown more used to the topic over time: the warnings of "time running out"... have been around for some time now.'

He said it was important to distinguish between whether people were personally concerned and whether they thought it was an important issue at a national level. The survey found that a 'surprisingly high proportion of people' named climate change as one of the major national concerns, he said.

The researchers found that people who had been directly affected by the flooding in the Somerset Levels were 'even more convinced of the reality and seriousness of climate change'.

A separate survey of 135 people whose properties were damaged by the flooding found higher levels of concern than among the general population, at 78 per cent.

Professor Nick Pidgeon of Cardiff University and lead researcher, said: 'Our findings demonstrate that an association between last year's winter flooding and climate change has been forming in the minds of many ordinary people in Britain, who also view these events as a sign of things to come.'

29 January 2015

⇨ The above information is reprinted with kind permission from *The Telegraph*. Please visit www.telegraph.co.uk for further information.

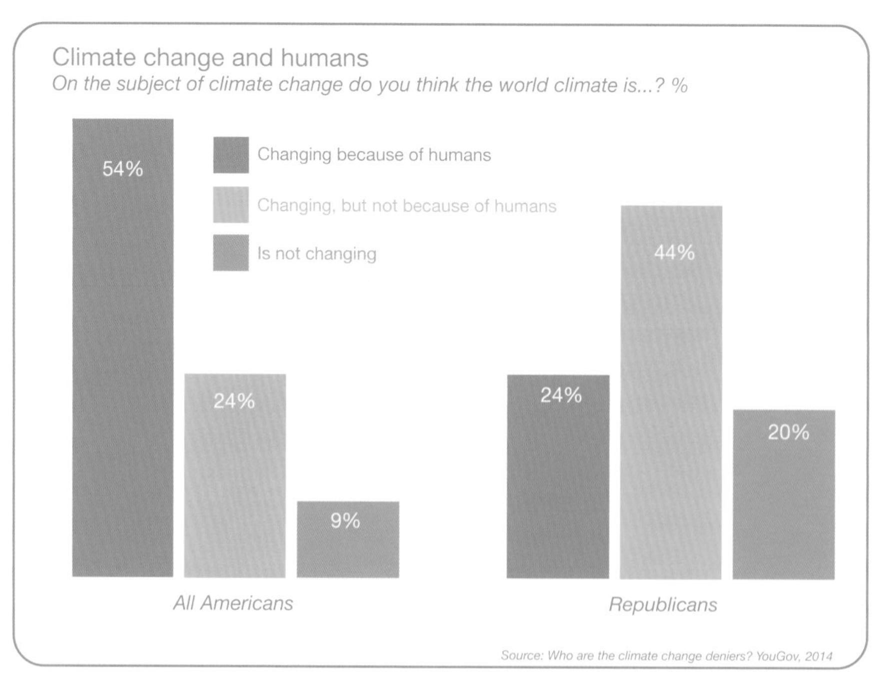

Policies and solutions

Ten action areas for growth

Implementing these actions will in many cases require significant investment. International and national public finance will be needed to catalyse and help leverage private finance, in particular for low-carbon energy and urban development; action to halt deforestation and restore degrade land; to build capacity; and to scale up research, development and demonstration of clean technologies and processes. The economic benefits of such investment will be substantial, even without consideration of the gains for the climate.

The Global Commission urges the international community to seize the opportunity of the unique series of meetings occurring in 2015 to put the world on a pathway to low-carbon, climate-resilient growth and development. Cooperative action, between governments at all levels and with the private sector, international organisations and civil society, can help achieve both better growth and a better climate. This will require strong and sustained political leadership. But the prize is immense. Together, a secure, prosperous and sustainable future is within our reach.

The Commission makes the following recommendations:

1. Accelerate low-carbon development in the world's cities

All cities should commit to developing and implementing low-carbon urban development strategies by 2020, using where possible the framework of the Compact of Mayors, prioritising policies and investments in public, non-motorised and low-emission transport, building efficiency, renewable energy and efficient waste management.

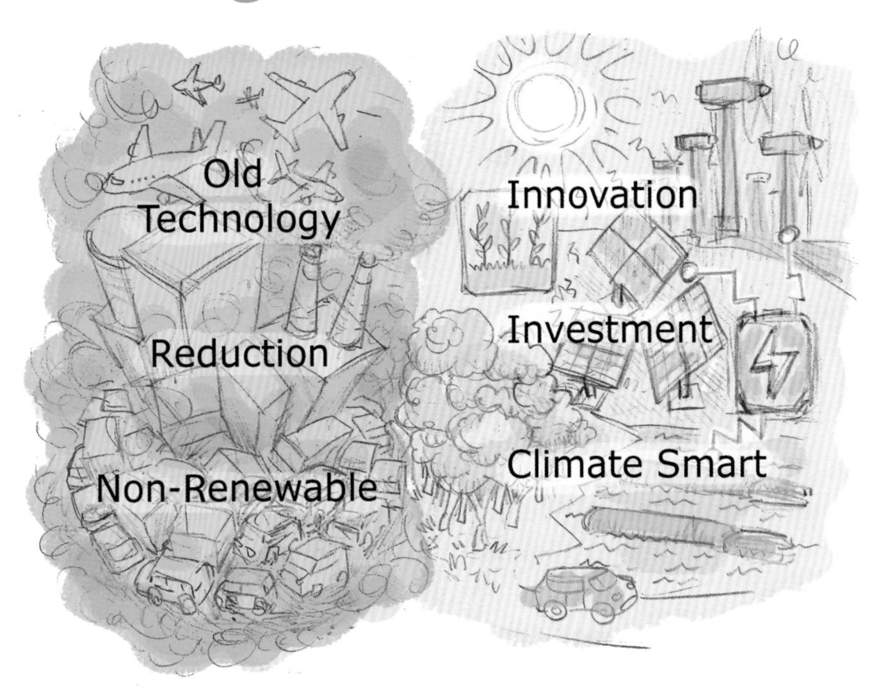

Compact, connected and efficient cities can generate stronger growth and job creation, alleviate poverty and reduce investment costs, as well as improving quality of life through lower air pollution and traffic congestion. Better, more resilient models of urban development are particularly critical for rapidly urbanising cities in the developing world. International city networks, such as the C40 Cities Climate Leadership Group, ICLEI (Local Governments for Sustainability) and United Cities and Local Governments (UCLG), are scaling up the sharing of best practices and developing initiatives to facilitate new flows of finance, enabling more ambitious action on climate change. Multilateral development banks, donors and others should develop an integrated package of at least US$1 billion for technical assistance, capacity building and finance to support commitments by the world's largest 500 cities. Altogether, low-carbon urban actions available today could generate a stream of savings in the period to 2050 with a current value of US$16.6 trillion, and could reduce annual GHG emissions by 3.7 Gt CO_2e by 2030.

2. Restore and protect agricultural and forest landscapes and increase agricultural productivity

Governments, multilateral and bilateral finance institutions, the private sector and willing investors should work together to scale-up sustainable land use financing, towards a global target of halting deforestation and putting into restoration at least 500 million ha of degraded farmlands and forests by 2030. Developed economies and forested developing countries should enter into partnerships that scale-up international flows for REDD+, focused increasingly on mechanisms that generate verified emission reductions, with the aim of financing a further 1 Gt CO_2e per year from 2020 and beyond. The private sector should commit to extending deforestation-free supply chain commitments for key commodities and enhanced financing to support this.

Halting deforestation and restoring the estimated one-quarter of agricultural lands worldwide which are severely degraded can enhance agricultural productivity and resilience, strengthen food security, and improve livelihoods for agrarian and forest communities in developing countries. Developing countries, supported by international partnerships between governments, the private sector and community organisations, and initiatives such as the New York Declaration on Forests, REDD+, the 20x20 Initiative in Latin America, the Africa Climate-Smart Agriculture Alliance and the Global Alliance for Climate Smart Agriculture, are helping to improve enabling environments for forest protection and agricultural production, and reducing and sharing investment risk to facilitate larger financial flows. The Consumer Goods Forum and companies representing 90% of the global trade in palm oil have committed to deforestation-free supply chains by 2020, while major commodity traders and consumers are working to widen such pledges to other forest commodities. Enhancing such partnerships could enable a reduction in annual GHG emissions from land use of 3.3–9.0 Gt CO_2e by 2030.

3. Invest at least US$1 trillion a year in clean energy

To bring down the costs of financing clean energy and catalyse private investment, multilateral and national development banks should scale-up their collaboration with governments and the private sector, and their own capital commitments, with the aim of reaching a global total of at least US$1 trillion of investment per year in low-carbon power supply and (non-transport) energy efficiency by 2030.

The rapid scale-up of low-carbon energy sources and energy efficiency is essential to drive global growth, connect the estimated 1.3 billion people currently lacking access to electricity and the 2.7 billion who lack modern cooking facilities, and reduce fossil fuel-related air pollution. Increasing international financing for energy access is a key priority. International cooperation coordinated by development finance institutions is helping improve the risk-return profile of clean energy projects, particularly for renewables and energy efficiency, lowering the cost of capital for investment and increasing its supply. It is also starting to drive a shift in investments away from new coal-fired power and fossil fuel exploration; this needs to be accelerated, starting with developed and emerging economies. Scaling up clean energy financing to at least US$1 trillion a year could reduce annual GHG emissions in 2030 by 5.5–7.5 Gt CO_2e.

4. Raise energy efficiency standards to the global best

G20 and other countries should converge their energy efficiency standards in key sectors and product fields to the global best by 2025, and the G20 should establish a global platform for greater alignment and continuous improvement of standards.

Co-operation to raise energy efficiency standards for appliances, lighting, vehicles, buildings and industrial equipment can unlock energy and cost savings, expand global markets, reduce non-tariff barriers to trade, and reduce air pollution and GHG emissions. Cooperation should be facilitated and supported by the G20, empowering existing sectoral initiatives, and international organisations such as the International Energy Agency (IEA), the International Partnership for Energy Efficiency Cooperation (IPEEC), and Sustainable Energy for All (SE4All). Globally, enhanced energy efficiency investments could boost cumulative economic output by US$18 trillion to 2035, increasing growth by 0.25–1.1% per year. Aligning and gradually raising national efficiency standards could reduce annual GHG emissions in 2030 by 4.5–6.9 Gt CO_2e.

5. Implement effective carbon pricing

All developed and emerging economies, and others where possible, should commit to introducing or strengthening carbon pricing by 2020, and should phase out fossil fuel subsidies.

Strong, predictable and rising carbon prices send an important signal to help guide consumption choices and investments in infrastructure and innovation; the fiscal revenues generated can be used to support low-income households, offset reductions in other taxes, or for other policy objectives. An estimated 12% of annual GHG emissions are now covered by existing or planned carbon taxes or trading systems around the world. Businesses are increasingly calling on governments to implement carbon pricing, and over 150 now use an internal carbon price (typically around US$40/t CO_2 for oil companies) to guide investment decisions. International cooperation on carbon pricing and subsidy reform, including through the G20 and with the support of the World Bank, the Organisation for Economic Co-operation and Development (OECD) and the International Monetary Fund (IMF), can help mitigate concerns about competitiveness impacts from unilateral policy measures, improve knowledge-sharing and transparency, provide opportunities to link emission trading schemes, and reduce the costs of action.

6. Ensure new infrastructure is climate-smart

G20 and other countries should adopt key principles ensuring the integration of climate risk and climate objectives in national infrastructure policies and plans. These principles should be included in the G20 Global Infrastructure Initiative, as well as used to guide the investment strategies of public and private finance institutions, particularly multilateral and national development banks.

About US$90 trillion in infrastructure investment is needed globally by 2030 to achieve global growth expectations, most of it in developing countries. Infrastructure investment has become a core focus of international economic cooperation through the G20 and for established and new development finance institutions. Integrating climate objectives into infrastructure decisions, often at no or very modest additional cost, will increase climate resilience and avoid locking in carbon-intensive and polluting investments. International finance will have to be significantly scaled up to deliver the up-front infrastructure investments needed to achieve development and climate goals, including increased capitalisation of both national and multilateral development banks.

7. Galvanise low-carbon innovation

Emerging and developed country governments should work together, and with the private sector and developing countries, in strategic partnerships to accelerate research, development and demonstration (RD&D) in low-carbon technology areas critical to post-2030 growth and emissions reduction.

Public funding for low-carbon RD&D is currently too low to catalyse innovation for long-term growth and cost-effective emissions reduction beyond 2030. It should be at least tripled by the major economies by the mid-2020s. International partnerships enable countries to share the costs of innovation, and the knowledge generated by it. This can be of particular benefit to low and middle-income countries, enabling them to 'leapfrog' to new technologies and enhance their innovation capacity. Priority areas for low-carbon cooperative innovation include agriculture and energy access, particularly in developing countries; longer-term global solutions such as bioenergy and carbon capture, utilisation and storage; and key technologies to avoid lock-in of carbon-intensive infrastructure, including buildings, electricity networks and transport systems.

8. Drive low-carbon growth through business and investor action

All major businesses should adopt short- and long-term emissions reduction targets and implement corresponding action plans, and all major industry sectors and value chains should agree on market transformation roadmaps, consistent with the long-term decarbonisation of the global economy. Financial sector regulators and shareholders should actively encourage companies and financial institutions to disclose critical carbon and environmental, social and governance factors and incorporate them in risk analysis, business models and investment decision-making.

Businesses are driving a US$5.5 trillion global market in low-carbon and environmental technologies and products, and many large companies are now cutting their emissions, realising significant cost savings and often enhancing profitability. Business- and finance sector-led initiatives are setting new norms for corporate action, including long-term target-setting and the integration of climate risk into investors' analysis and strategy. Initiatives such as the Tropical Forest Alliance 2020 and the Low Carbon Technology Partnership initiatives seek to transform markets in key sectors and value chains, driving innovation and creating global low-carbon markets. Companies should work with governments, unions and other stakeholders to ensure a 'just transition' to a low-carbon economy, supporting job creation, skills development and community renewal.

9. Raise ambition to reduce international aviation and maritime emissions

Emissions from the international aviation and maritime sectors should be reduced in line with a 2°C pathway through action under the International Civil Aviation Organization (ICAO) to implement a market-based measure and aircraft efficiency standard, and through strong shipping fuel efficiency standards under the International Maritime Organization (IMO).

Global aviation and shipping together produced about 5% of global CO_2 emissions, and by 2050 this is expected to rise to 10–32%. Yet they offer some of the most cost-effective emission reductions available today, particularly through improved fuel efficiency. Two new IMO standards are expected to save an average of US$200 billion in annual fuel costs by 2030. Adoption by the ICAO in 2016 of a market-based measure (an emissions trading or offset scheme) can both cut emissions and potentially generate finance for climate action or other purposes. This should be complemented by a new aircraft standard to ensure emissions reductions within the sector. The IMO should adopt a global emissions reduction target and promote fuel saving through strong operational efficiency standards and a supporting data-sharing system. These measures could help reduce annual GHG emissions by 0.6–0.9 Gt CO_2e by 2030.

10. Phase down the use of hydrofluorocarbons (HFCS)

Hydrofluorocarbons, used as refrigerants, as solvents, in fire protection and in insulating foams, are the fastest-growing GHGs in much of the world, increasing at a rate of 10–15% per year. Replacing HFCs with greener refrigerants has low upfront costs and can result in both energy and cost savings. Cooperative initiatives such as through the Climate and Clean Air Coalition to Reduce Short-Lived Climate Pollutants (CCAC), the Consumer Goods Forum, and Refrigerants, Naturally! are helping countries and companies scale back HFC use. Incorporating HFCs into the Montreal Protocol could realise significant near-term gains to slow climate change and provide support to developing countries, avoiding 1.1–1.7 Gt CO_2e of GHG emissions per year by 2030, while driving significant energy efficiency improvements.

⇨ The above information is reprinted with kind permission from The New Climate Economy. Please visit www.newclimateeconomy.net for further information.

UN and climate change: towards a climate agreement

Background

Climate change is a complex problem, which, although environmental in nature, has consequences for all spheres of existence on our planet. It either impacts on – or is impacted by – global issues, including poverty, economic development, population growth, sustainable development and resource management.

At the very heart of the response to climate change, however, lies the need to reduce emissions. In 2010, governments agreed that emissions need to be reduced so that global temperature increases are limited to below 2 degrees Celsius.

UN Framework Convention on Climate Change

In 1992, countries joined an international treaty, the United Nations Framework Convention on Climate Change (UNFCCC), to consider what they could do to limit global temperature increases and the resulting climate change, and to cope with its impacts.

By 1995, countries realised that emission reductions provisions in the Convention were inadequate. As a result, they launched negotiations to strengthen the global response to climate change, and, in 1997, adopted the Kyoto Protocol.

The Kyoto Protocol legally binds developed countries to emission reduction targets. The Protocol's first commitment period started in 2008 and ended in 2012. The second commitment period began on 1 January 2013 and will end in 2020.

There are now 195 Parties to the Convention and 192 Parties to the Kyoto Protocol. The Protocol entered into force on 16 February 2005. Since then, the Protocol the Parties to the Protocol have continued the negotiations and have amended the Protocol to achieve more ambitious results by 2030.

The following timeline provides a brief summary of negotiations towards a climate agreement.

Negotiations timeline

1979 – The first World Climate Conference takes place.

1988 – The Intergovernmental Panel on Climate Change (IPCC) is set up. Learn more about the science of climate change.

1990 – The IPCC and the second World Climate Conference call for a global treaty on climate change. The United Nations General Assembly negotiations on a framework convention begin.

1991 – First meeting of the Intergovernmental Negotiating Committee takes place.

1992 – At the Earth Summit in Rio, the UNFCCC is opened for signature along with its sister Rio Conventions, the UN Convention on Biological Diversity and the UN Convention to Combat Desertification.

1994 – The UNFCCC enters into force.

1995 – The first Conference of the Parties (COP 1) takes place in Berlin.

1996 – The UNFCCC Secretariat is set up to support action under the Convention.

1997 – The Kyoto Protocol is formally adopted in December at COP3.

2001 – The Marrakesh Accords are adopted at COP7, detailing the rules for implementation of the Kyoto Protocol, setting up new funding and planning instruments for adaptation, and establishing a technology transfer framework.

2005 – Entry into force of the Kyoto Protocol. The first Meeting of the Parties to the Kyoto Protocol (MOP 1) takes place in Montreal. In accordance with Kyoto Protocol requirements, Parties launched negotiations on the next phase of the KP under the Ad Hoc Working Group on Further Commitments for Annex I Parties under the Kyoto Protocol (AWG-KP). What was to become the Nairobi Work Programme on Adaptation (it would receive its name in 2006, one year later) is accepted and agreed on.

2007 – The IPCC's Fourth Assessment Report is released. Climate science entered into popular consciousness. At COP13, Parties agreed on the Bali Road Map, which charted the way towards a post-2012 outcome in two work streams: the AWG-KP, and another under the Convention, known as the Ad-Hoc Working Group on Long-Term Cooperative Action Under the Convention.

2009 – Copenhagen Accord drafted at COP15 in Copenhagen. Countries later submitted emissions reductions pledges or mitigation action pledges, all non-binding.

2010 – Cancun Agreements drafted and largely accepted by the COP, at COP16. Through the Agreements, countries made their emission reduction pledges official, in what was the largest collective effort the world has ever seen to reduce emissions in a mutually accountable way.

2011 – The Durban Platform for Enhanced Action drafted and accepted by the COP, at COP17. In Durban, governments clearly recognised the need to draw up the blueprint for a fresh universal, legal agreement to deal with climate change beyond 2020, where all will play their part to the best of their ability and all will be able to reap the benefits of success together.

2012 – The Doha Amendment to the Kyoto Protocol is adopted by the CMP at CMP8. The amendment includes: new commitments for Annex I Parties to the Kyoto Protocol who agreed to take on commitments in a second commitment period from 1 January 2013 to 31 December 2020; a revised list of greenhouse gases to be reported on by Parties in the second commitment period; and amendments to several articles of the Kyoto Protocol pertaining to the first commitment period and which needed to be updated for the second commitment period.

2013 – Key decisions adopted at COP19/CMP9 include decisions on further advancing the Durban Platform, the Green Climate Fund and Long-Term Finance, the Warsaw Framework for REDD Plus and the Warsaw International Mechanism for Loss and Damage. More on the Warsaw Outcomes.

2014 – COP20 held in December in Lima, Peru.

2015 – COP21 or CMP11 will be held in Paris, France in December.

*Source: UNFCCC

⇨ From *Towards a climate agreement*, by The United Nations and Climate Change,

Key points and questions: IPCC working group 3 report on mitigating climate change

The UN's Inter-Governmental Panel on Climate Change (IPCC) published the third volume of its Fifth Assessment Report on Climate Change: Mitigation of Climate Change in Berlin, Germany on 12 April 2014. The report was finalised after a six-day meeting attended by delegates from over 100 countries and a number of the Report's expert authors.

The IPCC's first report on the *Physical Science Basis of Climate Change* was published in September 2013. It showed clearly that climate change is happening now and greenhouse gas emissions from human activities are the dominant cause. The second report on the *Impacts of Climate Change and Adaptation* to them was published in March 2014. This showed how climate change is already having an impact and warned of severe, pervasive, and irreversible impacts in the future if action was not taken to address emissions. The report also considered steps countries and regions could take to adapt to the impacts of climate change.

This third report looks at how to address the issues and risks identified in the first two reports by reducing those activities that contribute to human-induced climate change. It looks at current greenhouse gas emissions, the levels they will need to fall to in future, and how this can be achieved. The report recognises that climate change is a global problem and looks at the contribution all regions can make in tackling the problem.

This is the most significant report on the topic since the IPCC's Fourth Assessment Report in 2007. It has been prepared over the last four years by 235 experts from across the world who reached their conclusions by reviewing thousands of published research papers. It has undergone peer review by many other scientists, experts and by IPCC member governments. The thoroughness of the process is without parallel in terms of scope, rigour, transparency and level of government engagement.

1. What are the key headline findings from the report?

- Greenhouse gas emissions are still rising, and the rate of increase has itself been increasing – most of this increase is being driven by increasing global prosperity.
- On a business-as-usual pathway, global mean temperatures will increase by three to five degrees over pre-industrial levels by the end of the century.
- Staying under the two degree limit is possible but increasingly difficult – it will require a wide range of changes, including changes in technology, institutions and behaviours.
- Efforts to reduce emissions needs to take place across all sectors (e.g. energy, transport, agriculture) and all regions – reductions in demand for energy (through, for example, energy efficiency measures) can play a big part.
- Many countries already have policies in place to reduce emissions, but much more needs to be done – investment in clean technology needs to be massively scaled-up and mitigation policies need to be integrated into broader political considerations, such as growth, jobs and the environment.
- Dealing with climate change needs international action – this is a 'commons' problem and requires international cooperation to solve it.

2. How high are GHG emissions today and what are the sources?

The report shows clearly what GHG emissions are today. It reports that:

- Annual emissions, at around 50 billion tonnes of carbon dioxide equivalent, have never been higher.
- Fossil fuel combustion is the largest source of emissions (approximately two thirds), and the largest driver of growth, with other significant contributions from industry and agriculture.
- Population growth has an impact on emissions but the biggest driver of increasing emissions is the rise in global prosperity – as people across the world, particularly in Asia, move out of poverty their consumption of energy increases and with it their fossil fuel use also increases.
- In the future, without a rapid move to low-carbon technologies, global emissions are projected to increase further – this increase will be driven particularly by further increases in prosperity (rather than population) in developing countries.

3. How quickly do we have to reduce GHG emissions?

The report considers a range of different possible futures ('emissions scenarios') ranging from those where emissions grow rapidly throughout the century to those where rapid decarbonisation leads to a very quick decrease in GHG emissions. It shows that there are a number of different emissions scenarios which are 'more likely that not' to avoid exceeding the two-degree target – all these scenarios require global emissions to peak no later than 2030 and decline rapidly thereafter:

- In 2050 global emissions would need to be about half what they are today.
- 'Negative emissions' (i.e. removing CO_2 from the atmosphere through a variety of techniques) may be necessary in the second half of the century to stay beneath the two-degree target.

- Current global efforts to reduce emissions (the 'Cancún pledges') are consistent with avoiding exceeding a temperature increase of three degrees, but more needs to be done if to stay beneath the two-degree target.

4. Is it possible to reduce emissions to the levels required?

The report confirms that:

- Deep cuts in GHG emissions to limit warming to 2°C relative to pre-industrial levels remain possible, yet will entail challenging changes.
- In the last few years many renewable energy technologies have got cheaper and improved in performance. A growing number of renewable energy technologies are now mature enough to be delivered at scale.
- Demand reductions achieved through energy efficiency and behavioural change will play a key role in realising the necessary emissions reductions. Demand reduction reduces the need for spending on new power stations and the associated infrastructure.
- Decarbonisation of electricity, through renewables, nuclear and carbon capture and storage (CCS), will need to happen rapidly, with the majority of electricity provided by low-carbon sources by the middle of the century.
- There is high agreement that nuclear power can play a big role in reducing emissions. Insulation of existing buildings, and application of new technology in construction, can play a large role in reducing emissions from buildings. A range of approaches can reduce emissions from industry. The IPCC is increasingly optimistic about the ability for the transport sector to play a role in mitigation.
- Ambitious action to reduce emissions will have a range of additional benefits. It can help improve air quality and energy security as well as drive improvements in human health, ecosystems, resource availability and the resilience of the energy system.

5. What are the implications of the findings for the UK?

The report does not focus on individual countries. In the UK:

- We have significantly increased UK deployment of renewable electricity and are taking steps to build new nuclear power stations and carbon-capture and storage power stations – this will increase the generation of low-carbon electricity in the UK.
- We have achieved a lot on energy efficiency in the UK: Insulating millions of homes, improving business efficiency and shifting to more efficient vehicles. There is more to do, in particular through further efforts on insulation, and shifting to low-carbon transport.
- We are supporting progress internationally: By pressing the EU to move to a 30% emissions reduction target by 2020 and 50% by 2030; agreeing, as part of the EU, with other countries to enter the second commitment period of the Kyoto Protocol; by working on changing the global political conditions; and by working on the 2015 agreement.

As a leading industrialised nation it is right that the UK take a lead, alongside other industrialised countries, in delivering emissions reductions. However, we recognise that climate change is a problem that requires a global solution and we shall continue to focus on achieving a global deal.

6. What are the costs and benefits of mitigation?

- Ambitious mitigation will slightly slow the rate of global economic growth. To put this into context, the report shows that experts expect the global economy to grow by between 300% and 900% by 2100. Set against these growth levels, it is estimated that global consumption would be around 5% less than it would have been had mitigation not taken place.
- These estimates do not include the costs of failing to tackle climate change – the Working Group III report only considers the cost of action, not the cost of failure to act. A comparison between the costs of action and inaction will be presented in the synthesis report in October.
- It is also important to note that these costs do not include the benefits of climate action – cleaner air, healthier lifestyles and diverse energy supplies all offer economic benefits. Evidence suggests that the benefits from improved air quality alone are equal to the costs of taking action – that is, measures to reduce emissions pay for themselves even before the climate benefit is taken into account. Mitigation would avoid millions of premature deaths from poor air quality over the course of the century.

7. What has changed since the last report?

- Sadly, sufficient steps to tackle global greenhouse gas emissions have yet to be taken and as a result emissions have risen.
- From its last major review of six years ago, the report does provide a strengthened body of evidence on how emissions can be reduced (so-called 'emissions pathways') in order to avoid exceeding the two-degree limit.

8. Is this the IPCC's main report?

This is the third of three major IPCC reports in the 5th Assessment Report Series. All these reports will be combined into the synthesis report, to be published in October 2014. *Details on what the government is doing to tackle climate change can be found on GOV.UK** There is also a wealth of valuable information and handy tips on how to reduce your own energy consumption and lead a greener life on the Energy Saving Trust website.

13 April 2014

- The above information is reprinted with kind permission from the Department of Energy & Climate Change. Please visit www.gov.uk for further information.

UK pledges to help hardest hit by climate change

The UK has committed up to £720 million to the Green Climate Fund, joining other major economies to help developing countries adapt to climate change and go low-carbon.

The fund will target developing countries, including the most vulnerable, to help them adapt to the adverse effects of climate change and limit or reduce greenhouse gas emissions.

After the UK's contribution, the climate fund now stands at around $9 billion with 13 countries already pledging, including $3 billion from the US, $1.5 billion from Japan, $1 billion from Germany and $1 billion from France.

The UK money comes from existing funds earmarked for international climate work under the UK's commitment for 0.7% of gross national income to overseas development assistance.

These pledges should help add momentum to efforts to secure a global deal at the UNFCCC negotiations in Lima and in Paris next year.

Energy and Climate Change Secretary Ed Davey said:

'The poorest and most vulnerable on the planet are already suffering the effects of climate change and it's our moral duty to act.

'From protecting low-lying islands and coastal settlements from the impact of rising sea levels to helping farmers struggling with lower crop yields caused by the weather effects of climate change – British aid can save lives.

'Along with the UK, other major economies such as the US, Japan, Germany and France have made substantial pledges and I urge other countries to be equally ambitious.'

The UK has a seat on the fund's governing board whose 24 members, drawn equally from developing and developed countries, will decide on and supervise the fund's spending.

The World Bank will manage the fund's financial assets as interim trustee, subject to a review three years after the fund starts in 2015.

Notes for editors

- The UK is committing to fund 12 per cent of the GCF up to a maximum of £720 million
- The UK pledge is based on the assumption the Board concludes the work it has already commissioned to ensure key operational elements of the Fund are in place, so it can start supporting projects as quickly as possible. These include accrediting implementing entities, agreeing the investment criteria, and the Private Sector Facility Business Plan.

Background of the GCF can be found on the GCF website in summary:

- The Fund will contribute to the achievement of the ultimate objective of the United Nations Framework Convention on Climate Change (UNFCCC). In the context of sustainable development, the Fund will promote the paradigm shift towards low emission and climate-resilient development pathways by providing support to developing countries to limit or reduce their greenhouse gas emissions and to adapt to the impacts of climate change, taking into account the needs of those developing countries particularly vulnerable to the adverse effects of climate change.
- The Fund aims to spend 50% on mitigation, and 50% on adaptation (with 50% of the adaption fund going to the most vulnerable)
- The Fund aims to start programming in 2015.

20 November 2014

- The above information is reprinted with kind permission from the Department of Energy & Climate Change. Please visit www.gov.uk for further information.

UN climate talks increasingly favour people alive today over future generations

THE CONVERSATION

***An article from* The Conversation.**

By Myles Allen, Professor of Geosystem Science, Leader of ECI Climate Research Programme at University of Oxford

If you ask the climate negotiators gathering in Bonn this week for their last get-together before the Paris conference in December who they are doing all this for, the reply would probably mention future generations. You can be as cynical as you like about what actually drives them but most people, including the negotiators themselves, no doubt think that the point of the exercise is to protect the interests of the planet as a whole.

But is it? And whose planet are they trying to protect? Worryingly, the basic negotiating framework of the UN climate process increasingly favours the interests of people alive today over future generations, in ways that perhaps even the negotiators themselves do not understand.

At the heart of the problem is the challenge of setting priorities between two very different kinds of climate pollutant. On the one hand, we have cumulative pollutants such as carbon dioxide, which build up in the climate system like heavy metals in the food chain. On the other, we have short-lived pollutants like methane and soot, which are 'washed out' naturally within a few weeks or years.

Action on short-lived pollutants has become very fashionable. It has an entire movement devoted to it, the Climate and Clear Air Coalition, enthusiastically supported by the United Nations Environment Programme. The reasons are clear: unlike carbon dioxide, many of these emissions can be reduced cheaply, with massive immediate benefits to human health and agriculture in precisely the countries where these emissions come from.

Who could possibly object to measures that would save the lungs and lives of women in developing countries and at the same time could help improve our prospects of keeping global temperatures below two degrees?

Look to the longterm

But there is a problem. As I explain in a new report released by the Oxford Martin School, as long as carbon dioxide emissions are still rising – and last year's blip notwithstanding, they are – emissions of short-lived climate pollutants are almost entirely irrelevant to peak warming.

The reason is simple: because carbon dioxide accumulates in the system, stabilising temperatures at any level means we have to get net global carbon dioxide emissions to zero. Even on the most heroic assumptions about future reductions, until carbon dioxide emissions are falling, and falling fast, net zero is still many decades off – by which time today's methane and soot emissions will have long since washed out of the climate system.

So why are countries so enthusiastic about them? Part of the reason is obscurely technical. For reasons long since forgotten, the whole UN process is based on the notion of 'carbon dioxide equivalent' emissions, with equivalence measured in terms of a kind of 'exchange rate' called the '100-year global warming potential'.

Given the name, you might be forgiven for thinking this was something to do with global warming over 100 years, but it isn't. It turns out GWP100 actually measures the relative impact of different emissions on warming rates over the next 30-40 years. And on this timescale, cutting short-lived climate pollutants could indeed have some impact.

Halving global methane emissions immediately, for example, could cut expected global temperatures by a couple of tenths of a degree by 2050 – which would be useful, but only comparable to natural fluctuations on these time scales. Halving global CO_2 emissions would have a much bigger impact (and an immeasurably bigger impact after 2050), but would also be far, far more difficult and expensive.

So much of the UN process is constructed around a measure of the impact of different emissions that explicitly focuses on the interests of the generation of decision-makers alive today. To be fair, some countries, like Brazil, have been calling for years for GWP100 to be replaced. But just replacing it with another exchange rate won't help, because any rate that works on one time scale will fail on another.

Time to stop pretending

The solution is extremely simple. Countries need to acknowledge the need to get net carbon dioxide emissions to zero and limit the total amount we dump in the atmosphere in the meantime. So until CO_2 emissions are falling, and falling fast enough that there is a realistic prospect of getting them to zero in the foreseeable future, we avoid the temptation of pretending that action on short-lived climate pollutants is helping to limit peak warming.

This issue arouses strong passions: colleagues and I were recently accused by fellow scientists of putting 'tens of millions of lives' at risk by calling for a delay in including short-lived climate pollutants in the UN climate process.

Let me be clear: I am all in favour of cutting soot emissions in developing countries. My first job, back in the 1980s, was developing clean and efficient wood stoves in East Africa. I well recall the infernal conditions experienced by women cooking over open fires in 40-degree heat: we don't need a global climate treaty to need a reason to do something about this kind of thing.

But until carbon dioxide emissions are falling, we shouldn't pretend cutting soot emissions is helping to stabilise global temperatures, because it isn't.

So a crucial test of whether the UN climate process is actually working in the interests of future generations is whether the negotiators in Bonn, and in Paris in December, acknowledge the need for net zero carbon dioxide emissions. Watch this space.

2 June 2015

⇨ The above information is reprinted with kind permission from The Conversation Trust (UK) Limited. Please visit www.theconversation.com for further information.

Geoengineering: the ethical problems with cleaning the air

Intervening in climate change currently raises more questions than answers when it comes to manipulating.

Last August, Berlin hosted an international conference on the controversial issue of geoengineering. We ask expert Dr Naomi Vaughan from the Tyndall Centre for Climate Change Research at the University of East Anglia to explain why it's such a hot potato for climate scientists.

What do we mean by geoengineering?

Geoengineering means large-scale interventions in the Earth's climate system to try to tackle climate change. There are, broadly, two types. The first would try to take some of the greenhouse gases that are causing climate change out of the atmosphere. The second would try to balance out that we have too many greenhouse gases in our atmosphere by reflecting more sunlight back into space.

How would we capture CO_2?

Oceans and land take a lot of carbon out of the atmosphere each year. There are ideas to try to make those processes happen more quickly [for example, by planting more trees]. The other type would be to use chemical engineering to scrub the atmosphere of CO_2 – but once you have caught it you need to store it somewhere for a long time.

And reflecting the sunlight?

The idea that would seem to be able to do a big enough job for us, given the challenge of climate change, is to reflect more sunlight back into space by putting tiny little particles high up into our atmosphere, into the stratosphere.

Have there been any experiments carried out?

There has been some laboratory-based work on the particle, what size the particle might be, what it might be made of and how it goes about reflecting the sunlight, but most of the work on reflecting sunlight back into space has been done on computer models or desk-based research. There are a couple of start-up companies looking at what chemical processes might be the best ones for capturing CO_2. [For] storage there are already demonstration projects globally.

What are the concerns with geoengineering?

The idea of reflecting more sunlight back into space raises a lot of ethical, moral and legal questions. If you do [use particles] you have to keep doing it. If you stop it, or it fails, the climate will bounce back but very quickly – quicker than the climate change that would have occurred. That causes a lot of concerns. You'd be affecting the global climate with an intentional intervention – so who is responsible for that? What if there are side effects? Taking CO_2 out of the atmosphere raises [fewer] big ethical questions, but there [is] still a set of questions about what we do with all this CO_2 we are going to store... what kind of burden are we placing on future generations?

How does it fit in with reducing emissions?

Reflecting sunlight back into space is not a replacement for reducing our emissions – it might be something we could do on top of reducing our emissions and adapting to changes. When we are talking about taking CO_2 out of the atmosphere it is a lot cheaper not to put it there in the first place.

6 November 2014

⇨ The above information is reprinted with kind permission from *The Guardian*. Please visit www.theguardian.com for further information.

Fighting desertification will reduce the costs of climate change

We should 'climate proof' the land to prevent some of the tragedies of a warmer world, says UN desertification chief.

By Monique Barbut

In the last five years, we have seen the disastrous impacts of droughts, heatwaves, cyclones, floods and flashfloods.

From the Philippines, Pakistan and the United States to the Middle East and Horn of Africa; death, food shortages, loss of homes and incomes or damage to infrastructure has followed in their wake. As disasters travel from one region to the next, many of us watch helplessly wondering what we can do about it.

Yet, everyone can do something that will make a difference. The battle against climate change will be won or lost at the local level, where you and I live.

Scientists agree that we can avoid catastrophe, if we stop emitting greenhouse gases today. However, we will still pay a price from the emissions we have produced so far. We know the impacts that different regions may face. What we do not know is where exactly the effects will show up and the extent of the damage.

Protecting the Earth

There is lots of evidence to show that the resilience of the land is a strong determinant of how much damage will be done.

This means we have a lot more power to protect ourselves than we thought or have acted upon. If we give due attention to the land in our neighbourhoods and local area we can take practical action that can reduce the costs of climate change to our family, community and the environment.

If we rehabilitate our soils so that they can filter and drain water better, we can reduce the occurrence of landslides, floods and flash floods. In regions where water is scarce, freshwater sources underground can recover and human, plant and animal migration can be reduced.

This land and ecosystem-based approach to adaptation is such a powerful tool for positive change. It makes us personally able to do something to avoid disaster. When our individual ecosystem-level initiatives are spread widely enough across countries, regions and the world, they will bring about a global transformation from the ground up.

Ecosystem-based adaptation demands that we give attention to the management and in many cases the recovery of natural resources. This will reduce the potential for disaster and help us secure more food, energy and water for our everyday needs.

But I emphasise paying attention to the land because it is often the forgotten natural resource.

Each of us can do something at the local level that can make the difference. Let us 'climate proof' the land. Let us start today, during the observance of the World Day to Combat Desertification, to strengthen the health and well-being of natural resources to make sure the worst human suffering and economic losses of climate change are avoided.

We may not eliminate all the impacts of climate change but with healthy and productive land many tragedies are preventable.

25 July 2014

⇨ The above information is reprinted with kind permission from Responding to Climate Change (RTCC). Please visit www.rtcc.org for further information.

How forestry helps address climate change

Forests help us address climate change by reducing the amount of greenhouse gases in the atmosphere.

They do this by absorbing carbon dioxide (CO_2), using the carbon (C) to produce sugars for tree growth and releasing the oxygen (O_2) back into the air. As trees grow, they store carbon in their leaves, twigs and trunk, and in the soil around them.

Globally, we have the capacity to increase the amount of carbon stored by forests by reducing the amount of deforestation in developing countries, and by converting non-forested areas to forest. Deforestation caused by the unsustainable harvesting of timber and the conversion of forests to other land uses, accounts for almost 20 per cent of global carbon dioxide emissions.

The forests and woodlands in Britain have a role to play too. They can be managed as a sustainable source of wood – an alternative and less polluting energy source to fossil fuels, and a low-energy construction material.

There are six key actions that should be taken now to protect what we have, and to make sure we can adapt to the new threats and opportunities that climate change will bring while still maintaining and expanding a sustainable forest and woodland resource.

Protect what we already have

- Reduce deforestation
- Restore the world's forest cover
- Use wood for energy
- Replace other materials with wood.

Plan to adapt to our changing climate

If we get this right, the world's forests will contribute significantly to climate change mitigation. They will also benefit national economies and the well-being of current and future generations.

Key facts:

- climate change is a global problem
- carbon dioxide emissions are the main cause of climate change
- deforestation accounts for almost 20 per cent of annual global emissions of carbon dioxide.

Key message:

There are two ways to reduce the level of carbon dioxide in the atmosphere.

We can reduce the amount we produce or we can develop ways to capture and store it. Trees have the unique ability to do both.

18 December 2014

- The above information is reprinted with kind permission from the Forestry Commission. Please visit www.forestry.gov.uk for further information.

Why aren't more climate activists vegan?

By Chris Lang

This week I decided to go vegan. The decision came after many years of being a (sometimes meat-eating) vegetarian, or a (sometimes vegetarian) meat-eater.

There are many reasons behind my decision. I recently read Jonathan Safran Foer's book, *Eating Animals*, which I picked up almost by accident. Having read it, I can't justify eating animals on moral grounds, given the current way animals are farmed.

Then there's the impact of cattle ranching on the Amazon, with vast areas of forest destroyed. And the deforestation caused by soy plantations, grown to provide animal feed.

And then there's impact of livestock on the climate. In 2014, Professor Tim Benton, at the University of Leeds, told *The Guardian*, 'The biggest intervention people could make towards reducing their carbon footprints would not be to abandon cars, but to eat significantly less red meat.'

I do try to reduce my carbon footprint. I cycle or take public transport most of the time. I don't own a car. I've taken one short-haul flight in the last two years. I buy organic vegetables.

And every now and then, I hand over money to one of the most unpleasant, polluting and environmentally disastrous industries on the planet: the meat industry. Until Monday, I poured milk on my cornflakes every morning.

Last year Peter Scarborough and colleagues from the University of Oxford collected data on the real diets of more than 50,000 people in UK, including vegans, vegetarians, fish-eaters and meat-eaters. They analysed the greenhouse gas emissions associated with each diet.

In a paper published in *Climatic Change*, they reported that an average high meat diet has 2.5 times as many greenhouse gas emissions than an average vegan diet. Over a year, the saving in emissions is the equivalent of 1.5 tonnes of carbon dioxide.

What proportion of greenhouse gas emissions comes from the livestock sector?

In 2006, the FAO produced a report on the climate impacts of livestock. Titled *Livestock's long shadow*, the report estimated that the livestock sector accounts for 18% of global greenhouse gas emissions.

Seven years later, the FAO produced another report, *Tackling climate change through livestock*. This time, FAO reports that the livestock sector represents 14.5% of human-caused greenhouse gas emissions.

On its website, the FAO states that 'The two figures cannot be accurately compared, as reference periods and sources differ'. FAO experts told *The Guardian* that the new figure was based on 'a revised modelling framework and updated data, using new guidelines from the Intergovernmental Panel on Climate Change'.

In a 2009 article for *World Watch* magazine, Robert Goodland and Jeff Anhang estimate that livestock accounts for 51% of human-caused greenhouse gas emissions. Predictably, on its website PETA uses the 51% figure.

Regardless of the exact figure it's safe to say, as the FAO does in its 2013 report, that 'the livestock sector plays an important role in climate change'.

Reduce emissions from livestock, or reduce meat and dairy consumption?

The FAO argues that a 30% reduction of greenhouse gas emissions is possible in the livestock sector by a range of methods, including increasing efficiency, improving manure management practices, sequestering carbon in grassland, and sourcing low emission intensity inputs.

The FAO notes the need for collective, concerted and global action:

'Recent years have seen interesting and promising initiatives by both the public and private sectors to address sustainability issues. Complementary multi-stakeholder action is required to design and implement cost-effective and equitable mitigation strategies, and to set up the necessary supporting policy and institutional frameworks.'

But neither the word 'vegetarian' nor the word 'vegan' appear anywhere in the FAO's 139-page-long report. (Neither does the word 'cruelty' make an appearance, although 'welfare' appears four times.) Instead, the FAO argues that world population will increase to 9.6 billion by 2050 and that,

'Driven by strong demand from an emerging global middle class, diets will become richer and increasingly diversified, and growth in animal-source foods will be particularly strong; the demand for meat and milk in 2050 is projected to grow by 73 and 58 per cent, respectively, from their levels in 2010.'

As well as working to reduce emissions from the meat and dairy industry, wouldn't it make sense to reduce consumption of meat? That's what a 2010 UNEP report suggested:

'A substantial reduction of impacts [from agriculture] would only be possible with a substantial worldwide diet change, away from animal products.

Of course, I'm not arguing for reductions in emissions from livestock so that we can continue to burn fossil fuels. To address climate change we need to leave fossil fuels in the ground, as well as reducing emissions from cow farts (and deforestation).

Dietary change is essential

Last year, Chatham House published a report looking at global public opinion on meat and dairy consumption. The report points out that 'consumption of meat and dairy produce is a major driver of climate change', and that 'shifting global demand for meat and dairy produce is central to achieving climate goals'.

The problem, as Rob Bailey, the report's lead author, told *The Guardian*, is that,

'A lot is being done on deforestation and transport, but there is a huge gap on the livestock sector. There is a deep reluctance to engage because of the received wisdom that it is not the place of governments or civil society to intrude into people's lives and tell them what to eat.'

Which is a good point. In September 2013, the German Green politician Renate Künast proposed the idea of one day a week vegetarian food in canteens. It became one of the most debated proposals from any party during the German election campaign, getting as much airtime as Syria and the eurozone combined.

The Greens were accused of pushing for an 'ecological dictatorship'. Rainer Brüderle, from the Free Democrats Party (FDP), then-junior coalition partners, said:

'People are smart enough to decide on their own when they eat meat and vegetables and when they don't. Constantly telling people what they do is not my understanding of freedom and liberty.'

But in its report, Chatham House argues that 'dietary change is essential if global warming is not to exceed two degrees Celsius'.

So how should climate activists communicate the importance of eating less meat? How come more people who are concerned about climate change aren't vegan?

And, come to think of it, what took me so long?

26 June 2015

⇨ The above information is reprinted with kind permission from REDD Monitor. Please visit www.redd-monitor.org for further information.

Key facts

- Since the last ice age, which ended about 11,000 years ago, Earth's climate has been relatively stable at about 14°C. However, in recent years, the average temperature has been increasing. (page 1)
- Scientific research shows that the climate – that is, the average temperature of the planet's surface – has risen by 0.89°C from 1901 to 2012. (page 1)
- Changes in the seasons (such as the UK spring starting earlier, autumn starting later) are bringing changes in the behaviour of species; for example, butterflies appearing earlier in the year and birds shifting their migration patterns. (page 1)
- Since 1900, sea levels have risen by about 10cm around the UK and about 19cm globally, on average. The rate of sea-level rise has increased in recent decades. (page 1)
- Arctic sea-ice has been declining since the late 1970s, reducing by about 4%, or 0.6 million square kilometres (an area about the size of Madagascar) per decade. At the same time Antarctic sea-ice has increased, but at a slower rate of about 1.5% per decade. (page 1)
- If global emissions are not reduced, average summer temperatures in the south east of England are projected to rise by over 2°C by the 2040s (hotter than the 2003 heatwave which was connected to 2,000 extra deaths in the UK). (page 3)
- These extreme weather events in the UK are likely to increase with rising temperatures, causing: heavier rainfall events – with increased risk of flooding. (page 3)
- Even with low levels of warming (less than 2°C above the temperature in 1800), global production of major crops such as wheat, rice and maize may be harmed. (page 4)
- The Climate Change Act became UK Law on 26 November 2008. This legislation introduced ambitious and legally binding national targets for the UK to reduce GHG emissions to 34% below base year by 2020 and to 80% below base year by 2050. These targets are underpinned with legally binding five-year GHG budgets. (page 7)
- Direct measurements of CO_2 in the atmosphere and in air trapped in ice show that atmospheric CO_2 increased by about 40% from 1800 to 2012. (page 7)
- New and resurgent vector-borne diseases, including Dengue, are moving north as temperatures rise, and have recently appeared in Europe. (page 9)
- Extreme weather conditions are projected to increase and are likely to hit UK food production. (Extreme weather led to the UK needing to import wheat in 2012 and 2013.) (page 9)
- Climate change poses an increasing risk to geopolitical stability and therefore the security and well-being of all. (page 10)
- The recent IPCC AR5 report concluded the climate is changing and there is a 95% certainty that it is caused by our actions – specifically the burning of fossil fuels, deforestation and land use change. (page 11)
- A recent assessment of 16,857 species by the International Union for the Conservation of Nature found that up to 9% of all birds, 15% of amphibians and 9% of corals are both highly vulnerable to climate change and already threatened with extinction. (page 14)
- In July 2010, Pakistan was affected by heavy monsoon rains, which led to massive flooding in the Indus River basin. More than ten million people were displaced, with about 20% of the country under water. The death toll was around 2,000. (page 17)
- In total, 79% of the British public think the world's climate is changing, however, 23% think it is not man-made, compared to 56% that believe it is. 7% believe the world's climate isn't changing and 14% don't know. (page 25)
- 39% believe that concerns over climate change have been exaggerated, although 47% believe the threat is as real as scientists have said. (page 25)
- The percentage of people unaware of the global phenomenon rises to more than 65% in developing countries such as Egypt, Bangladesh and India, whereas only 10% of the public is unaware in North America, Europe, and Japan. (page 25)
- In a paper published in Climatic Change, they reported that an average high meat diet has 2.5 times as many greenhouse gas emissions than an average vegan diet. Over a year, the saving in emissions is the equivalent of 1.5 tonnes of carbon dioxide. (page 38)

Adaptation

In relation to climate change, adaptation aims to respond to the effects of global warming by adapting to altered environments. This includes adapting to changed food production methods, agriculture and sea levels.

CO_2 emissions

Carbon dioxide gas released into the atmosphere. CO_2 is released when fossil fuels are burnt. An increase in CO_2 emissions due to human activity is arguably the main cause of global warming.

Carbon footprint

A carbon footprint is a measure of an individual's effect on the environment, taking into account all greenhouse gases that have been emitted for heating, lighting, transport, etc. throughout that individual's average day.

Carbon offsets

Carbon offsets are a reduction in greenhouse gas emissions made in order to compensate for greenhouse gas production somewhere else. Offsets can be purchased in order to comply with caps, such as the Kyoto Protocol. For example, rich industrialised countries may purchase carbon offsets from a developing country in order to satisfy environmental legislation.

Climate change

Climate change describes a global change in the balance of energy absorbed and emitted into the atmosphere. This imbalance can be triggered by natural or human processes. It can cause either regional or global changes in weather averages and frequency of severe climatic events.

Climate change refugees

Also known as 'environmental migrants', climate change refugees are people who have been forced to flee their home region following severe changes in their local environment as a result of global warming.

Climate models

Scientific models which are designed to replicate the Earth's climate. Scientists are able to hypothetically test the effects of global warming by simulating changes to the Earth's atmosphere.

Geoengineering

Geoengineering means large-scale interventions in the Earth's climate system to try to tackle climate change. There are, broadly, two types. The first would try to take some of the greenhouse gases that are causing climate change out of the atmosphere. The second would try to balance out that we have too many greenhouse gases in our atmosphere by reflecting more sunlight back into space.

Global warming

This refers to a rise in global average temperatures, caused by higher levels of greenhouse gases entering the atmosphere. Global warming is affecting the Earth in a number of ways, including melting the polar ice caps, which in turn is leading to rising sea levels.

Greenhouse gases (GHG)

A greenhouse gas is a type of gas that can absorb and emit longwave radiation within the atmosphere: for example, carbon dioxide, methane and nitrous oxide. Human activity is increasing the level of greenhouse gases in the atmosphere, causing the warming of the Earth. This is known as the greenhouse effect.

IPCC

An abbreviation for Intergovernmental Panel on Climate Change, the leading scientific body which assesses and reviews global climate change. It was founded by the United Nations Environment Programme and the World Meteorological Organization and currently has 194 member countries from around the world.

Kyoto Protocol

An international treaty setting binding targets for 37 developed countries to reduce their greenhouse gas emissions by at least five per cent below 1990 levels for the years 2008-2012. It was made international law in 2005. It was the world's first international agreement on tackling climate change.

Mitigation

In relation to climate change, mitigation refers to the act of reducing, or limiting, the level of greenhouse gas emissions in order to slow the rate of global warming. Emissions targets, government campaigns and the development of 'greener' energy sources are examples of how mitigation can be used to reduce climate change.

REDD

An abbreviation for the United Nations programme Reducing Emissions from Deforestation and Forest Degradation. This collaborative initiative aims to assist developing countries in combatting deforestation, illegal logging and fires in order to limit climate change.

Assignments

Brainstorming

- In small groups, discuss what you know about climate change. Consider the following points:
  - What is climate change?
  - What are the processes that cause climate change?
  - What is the IPCC and what does it do?

Research

- Design and present a ten-minute PowerPoint presentation on the predicted future impacts of climate change. Explain what is expected to happen to global temperatures and sea levels.
- What is your carbon footprint? Visit www.carboncalculator.direct.gov.uk to find out your carbon footprint. How could you reduce this? Compare your findings with others in your class. Is there much variation throughout your class or do you all have similar footprints?
- Do some research about geoengineering and write a report on your findings.
- Research climate change deniers and their viewpoints, then write an article or blog post exploring your findings.

Design

- Design a wall poster giving information on the Kyoto Protocol. Use a mind map or a spider diagram design to display key facts and figures about the Protocol. Include answers to questions such as: what is the Kyoto Protocol? When was did it become law? Who does it affect? Do you think it is successful in reducing greenhouse gas levels?
- Write an informative leaflet explaining what carbon offsets are and how they work. Include information on who might purchase carbon offsets, and how. You should make it readable and attractive, suitable for a teenage audience. You can add illustrations or diagrams to your leaflet to make it more accessible.
- Design a poster that will raise awareness of the impacts of climate change.
- Design a leaflet that explains climate change. Try to include some maps and statistics, as well as images or drawings.
- Choose one of the articles in this book and create an illustration to highlight the key themes/message of your chosen article.

Oral

- In small groups, role play a radio talk show on the topic of climate change. One student will play the radio show host, another a climate change expert who aims to dispel some common myths, and a third student will be a climate change sceptic arguing that human actions do not cause global warming. Other students can play listeners phoning in with questions. The host should aim to stimulate a lively debate on the topic, giving equal time to all arguments.
- Survey shows 88 per cent of the public believe climate is changing yet a record low of just 18 per cent are 'very concerned' about it. Why do you think this is? Discuss this issue as a class.
- There is already evidence that climate change is happening; however, there is some debate over the extent to which human activity is causing or exacerbating this. Using the articles concerning the climate debate, as well as your own research, summarise the arguments for and against the theory that human activity is the main reason for climate change. Create two lists, one supporting and one opposing the theory. Which list is longer? Which do you agree with? Discuss your findings with the class.

Reading/writing

- Climate change is a phrase used widely in modern society, particularly by the mass media and politicians, but what exactly is it? Write a brief summary of what climate change is and the processes which cause it.
- Watch Al Gore's 2006 documentary film *An Inconvenient Truth*. Do you find his arguments convincing? Do you think this film is an effective way of conveying the climate change message? Write a review of his presentation.
- Climate change solutions fall into two categories: adaptation and mitigation. Write an informative article explaining the differences between mitigation and adaptation. What are the advantages and disadvantages of each?
- Andrew Pendleton argues that 'Innovation, not emissions targets, should be the prime focus of international efforts.' Discuss this statement in groups. Do you agree with Pendleton? Do you think legislation can address climate change or should governments focus on developing 'greener technology'? Write a summary of your conclusions.
- Imagine you are a climate change refugee/ environmental migrant and have been forced out of your home due to the effects of climate change. Write a diary entry detailing why you were forced out of your home and your journey afterwards. How do you feel? Where will you go? What will you do? You may find reading *Moving stories* (page 16) helpful.

Index

adaptation measures 8, 9, 37
agriculture
  degraded farmlands, restoring 27–8
  greenhouse gas emissions 6, 39
  impact of climate change on 2–3, 8–9
Arctic communities, impact of climate change on 17
Arctic Paradox 12–13
Australia, heatwaves in 11–12
aviation emissions 29

business sector
  impact of climate change on 10
  low-carbon growth strategies 10, 29

carbon capture and storage (CCS) 33
carbon dioxide emissions 2, 6, 7, 8, 15, 19, 35, 38
  cumulative pollutant 35
  geoengineering strategies 36
  reduction strategies and targets 7, 33
carbon pricing 28
carbon sinks 6, 15, 19, 38
carbon-trading schemes 20
clean energy projects 20, 28, 33
climate change
  causes of 2, 19
  costs and benefits of mitigation 4, 20, 33
  definition 1
  effects of 2–4, 8–10
  evidence for 1, 2, 19–20
  future scenarios 2, 20
  human agency 2, 4, 7, 8, 11, 21, 23, 26
  natural cycles of 2, 23
  policies and solutions 20, 27–9
climate change sceptics 21, 22, 25
climate models 2, 7, 23
  flaws 18, 23
cooperative action to tackle climate change 4, 27–9
  see also Intergovernmental Panel on Climate Change (IPCC); United Nations Framework Convention on Climate Change (UNFCCC)
coral reefs 3, 14
crop yields 9
cumulative pollutants 35

decarbonisation 9, 33
deforestation 2, 11, 27, 28, 38
desertification, combating 37
diet, implications for greenhouse gas emissions 38–9
droughts 3, 8, 9, 20

ecosystems
  ecosystem-based adaptations 37, 38
  impact of climate change on 3, 8, 14–15, 16
energy efficiency 9, 28, 33
extinctions, species 3, 14
extreme weather events 3–4, 8, 9, 11–13, 14, 19–20

flooding 3, 8, 9, 12, 14, 16–17, 20, 26
food production, impact on 2–3, 9–10
forestry
  deforestation 2, 11, 27, 28, 38
  forest fires 14
  protecting 15, 38
fossil fuel combustion 2, 6, 11, 20, 32, 39
freezing weather 12, 13

geoengineering 36
geostability, threats to 10
glacial and interglacial periods 2, 23
glacier retreat 1, 3, 8, 16, 19
global warming
  20C pathway 5, 8, 9, 20, 30, 32
  see also climate change; temperature rises
Green Climate Fund 34
greenhouse effect 2
greenhouse gases 2, 6–7, 8, 19, 20, 32
  direct and indirect gases 6
  emissions scenarios 32–3
  geoengineering strategies 36
  reduction targets 7, 29, 33
  sources of 6
  see also carbon dioxide emissions; methane; water vapour

health, impact on 4, 9
heatwaves 3, 4, 8, 9, 11–12, 14, 20
human activity, and climate change 2, 4, 7, 8, 11, 21, 23, 26
hydrofluorocarbons (HFCs) 6, 29

ice sheets, shrinking 1, 5, 8, 15
industrial pollution 2, 6
infrastructure policies, climate-smart 28–9
Intergovernmental Panel on Climate Change (IPCC) 5, 14, 19, 20, 21, 24, 25, 26, 30, 32
  Assessment Reports 25, 31, 32–3

Kyoto Protocol 6–7, 30, 31, 33

land use change 11, 19, 20
livestock farming and greenhouse gas emissions 6, 39
low-carbon development strategies 27
low-carbon economy 29
low-carbon energy sources 6, 9, 28, 33
low-carbon innovation 29

maritime emissions 29
methane 2, 6, 7, 15–16, 19, 35

nuclear power 33

ocean acidification 3, 8, 14
ocean warming 2, 3, 15

Pakistan, floods in 16–17
permafrost, thawing of 5, 19
polar amplification of global warming 20
population movements, climate change and 17

poverty 4, 10, 34
prosperity, impact on greenhouse gas emissions 32
public awareness of climate change issues 25–6

rainfall patterns, changes in 1, 4, 12, 19–20
reflecting sunlight back into space (geoengineering strategy) 36

sea level rises 1, 3, 7, 8, 9, 13, 16, 17, 19, 20, 23
sea-ice, melting 1, 3, 12, 17, 19, 20, 23
seasons, changing patterns of 1
short-lived pollutants 35
snow cover, reduction in 19, 20
soot emissions 35–6
storms 12, 13

temperature rises 1, 2, 3, 4, 8, 14, 19, 20, 23, 32
  pauses in 8, 23, 24, 25
'tipping points' 4, 5
typhoons 13, 26

uncertainties in climate science 23, 24
United Nations Framework Convention on Climate Change (UNFCCC) 30–1, 34
urban development strategies 27

vegetarian and vegan diets 38, 39

waste disposal, and greenhouse gas emissions 6
water resources, impact of climate change on 3, 16
water vapour 19

Acknowledgements

The publisher is grateful for permission to reproduce the material in this book. While every care has been taken to trace and acknowledge copyright, the publisher tenders its apology for any accidental infringement or where copyright has proved untraceable. The publisher would be pleased to come to a suitable arrangement in any such case with the rightful owner.

Images

All images courtesy of iStock, except page 13 © NASA Goddard Space Flight Center.

Illustrations

Don Hatcher: pages 1 & 22, Simon Kneebone: pages 24 & 34, Angelo Madrid: pages 11 & 27.

Additional acknowledgements

Editorial on behalf of Independence Educational Publishers by Cara Acred.

With thanks to the Independence team: Mary Chapman, Sandra Dennis, Christina Hughes, Jackie Staines and Jan Sunderland.

Cara Acred

Cambridge

September 2015